A2 in a Week

Chemistry

Da[...]
Abbey College, Manchester
Series Editor: Kevin Byrne

D0244330

Where to find the information you need

Energetics 2	3
Thermodynamics	9
Chemical Equilibria 2	13
Acids and Bases	17
Kinetics 2	23
Redox Equilibria	28
Periodic Table 2	35
Transition Metals	42
Haloalkanes 2	49
Aromatic Compounds	54
Acids, Esters and Acid Chlorides	59
Carbonyls	64
Organic Nitrogen Compounds	68
Organic Spectroscopy	74
Synoptic Questions	80
Use your Knowledge Answers	84

SUCCESS OR YOUR MONEY BACK

Letts' market leading series A2 in a Week gives you everything you need for exam success. We're so confident that they're the best revision books you can buy that if you don't make the grade we will give you your money back!

HERE'S HOW IT WORKS

Register the Letts A2 in a Week guide you buy by writing to us within 28 days of purchase with the following information:

- Name
- Address
- Postcode
- Subject of A2 in a Week book bought

Please include your till receipt

To make a **claim**, compare your results to the grades below. If any of your grades qualify for a refund, make a claim by writing to us within 28 days of getting your results, enclosing a copy of your original exam slip. If you do not register, you won't be able to make a claim after you receive your results.

CLAIM IF...

You are an A2 (A Level) student and do not get grade E or above.
You are a Scottish Higher level student and do not get a grade C or above.
This offer is not open to Scottish students taking SCE Higher Grade, or Intermediate qualifications.

Registration and claim address:

Letts Success or Your Money Back Offer, Letts Educational, Chiswick Centre, 414 Chiswick High Road, London W4 5TF

TERMS AND CONDITIONS

1. Applies to the Letts A2 in a Week series only
2. Registration of purchases must be received by Letts Educational within 28 days of the purchase date
3. Registration must be accompanied by a valid till receipt
4. All money back claims must be received by Letts Educational within 28 days of receiving exam results
5. All claims must be accompanied by a letter stating the claim and a copy of the relevant exam results slip
6. Claims will be invalid if they do not match with the original registered subjects
7. Letts Educational reserves the right to seek confirmation of the level of entry of the claimant
8. Responsibility cannot be accepted for lost, delayed or damaged applications, or applications received outside of the stated registration/claim timescales
9. Proof of posting will not be accepted as proof of delivery
10. Offer only available to A2 students studying within the UK
11. SUCCESS OR YOUR MONEY BACK is promoted by Letts Educational, Chiswick Centre, 414 Chiswick High Road, London W4 5TF
12. Registration indicates a complete acceptance of these rules
13. Illegible entries will be disqualified
14. In all matters, the decision of Letts Educational will be final and no correspondence will be entered into

Letts Educational
Chiswick Centre
414 Chiswick High Road
London W4 5TF

Tel: 020 8996 3333
Fax: 020 8742 8390
e-mail: mail@lettsed.co.uk
website: www.letts-education.com

Every effort has been made to trace copyright holders and obtain their permission for the use of copyright material. The authors and publishers will gladly receive information enabling them to rectify any error or omission in subsequent editions.

First published 2001

Text © Dan Evans and Alex Watts 2001
Design and illustration © Letts Educational Ltd 2001

British Library Cataloguing in Publication Data
A CIP record for this book is available from the British Library.

ISBN 1 85805 917 8

Prepared by *specialist* publishing services, Milton Keynes
Typesetting and artwork by Pantek Arts Ltd, Maidstone, Kent.
Printed in Italy

Letts Educational Limited is a division of Granada Learning Limited, part of the Granada Media Group

Energetics 2

10 minutes

Test your knowledge

1 Write an equation to represent each of the following enthalpy changes:
   a) Standard enthalpy of atomisation of carbon.
   b) Standard enthalpy of atomisation of oxygen.
   c) Standard enthalpy of hydration of sodium ions.
   d) Standard enthalpy of hydration of chloride ions.
   e) Lattice enthalpy of sodium chloride.

2 Using the energy values below construct a Born–Haber cycle and use it to determine the lattice enthalpy for lithium chloride.

$\Delta H_{f(LiCl)} = -409$; $\Delta H_{at(Li)} = 159$; First ionisation energy of lithium = 520

$\Delta H_{at(Cl_2)} = 122$; First electron affinity of chlorine = −349

3 Name the two factors that influence the value of the lattice enthalpy of a compound.

4 The variation in solubility of the Group 2 sulphates and hydroxides can be understood by a consideration of which two enthalpy changes?

Answers

4 Lattice enthalpy and hydration enthalpy.

3 Radius and charge on the ions.

2 −861 kJmol⁻¹

1 a) $C_{(s)} \rightarrow C_{(g)}$ b) $\frac{1}{2}O_{2(g)} \rightarrow O_{(g)}$ c) $Na^+_{(g)} + aq \rightarrow Na^+_{(aq)}$
   d) $Cl^-_{(g)} + aq \rightarrow Cl^-_{(aq)}$ e) $Na^+_{(g)} + Cl^-_{(g)} \rightarrow Na^+Cl^-_{(s)}$

 If you got them all right, skip to page 7

Energetics 2

10 minutes

Improve your knowledge

1 The enthalpy of **atomisation** ($\Delta H^\ominus_a$) is the enthalpy change when one mole of isolated gaseous atoms is formed from the element under standard conditions. The equation that represents the enthalpy of atomisation of chlorine is:

$$^1/_2\ Cl_{2(g)} \rightarrow Cl_{(g)}$$

The enthalpy of **hydration** ($\Delta H^\ominus_{hyd}$) is the enthalpy change when one mole of gaseous ions is dissolved in excess water under standard conditions. The equation that represents the enthalpy of hydration of sodium ions is:

$$Na^+_{(g)} + aq \rightarrow Na^+_{(aq)}$$

The **lattice enthalpy** ($\Delta H^\ominus_{latt}$) of an ionic substance is the enthalpy change when one mole of an ionic substance is formed from its constituent gaseous ions. The general equation for the lattice enthalpy of a substance is:

$$M^+_{(g)} + X^-_{(g)} \rightarrow M^+X^-_{(s)}$$

2 A **Born–Haber cycle** is an extension of Hess's law and enables us to calculate lattice enthalpies of ionic compounds which are difficult to determine experimentally.

By considering all of the associated energy changes in forming an ionic lattice the lattice enthalpy can be calculated.

Using the following data a Born–Haber cycle can be drawn to determine the lattice enthalpy of sodium chloride.

$\Delta H_{f\ (NaCl\ (s))} = -411$ kJ mol^{-1} ($\Delta H_{f\ (NaCl\ (s))}$ represents the standard enthalpy of formation of sodium chloride)

$\Delta H_{at\ (Na\ (s))} = +108.4$ kJ mol^{-1} (at = atomisation)
first ionisation energy of sodium = +500 kJ mol^{-1}

$\Delta H_{at\ (Cl_2\ (g))} = +121.1$ kJ mol^{-1} (This forms one mole of isolated chlorine atoms i.e. $^1/_2\ Cl_2 \rightarrow Cl$)
First electron affinity for chlorine = -364 kJ mol^{-1}

Key points from AS in a Week
Energetics 1
pages 34–35
Lattice/hydration enthalpy
pages 18–19

Plotting this data enables the lattice energy to be determined.

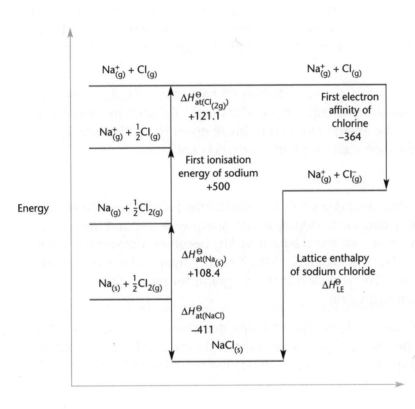

The lattice enthalpy is calculated to be -776 kJ mol^{-1}

Values determined from a Born–Haber cycle may differ from experimentally determined values. This is because a Born–Haber cycle assumes 100% ionic character and substances considered as ionic always have a degree of covalent character.

The **lattice enthalpy** of an ionic compound is the enthalpy change when **one mole** is separated into its constituent ions under **standard conditions**.

3 The **radius** and **charge** of ions affect the value of the lattice enthalpy. The smaller the ion, the greater the charge density of the ion and so the more it will attract an ion of the opposite charge, hence the greater the value of the lattice enthalpy. Similarly, ions with a higher charge will also have a higher charge density.

The concepts of lattice enthalpy and hydration enthalpy can be used to explain the trend in solubilities of the Group 2 hydroxides and sulphates. In order for an ionic solid to dissolve, the energy required to break down the lattice (the lattice enthalpy) must be compensated for by the hydration enthalpy.

Considering the Group 2 hydroxides:

The lattice enthalpy decreases (becomes less exothermic) at a greater rate than the hydration enthalpy decreases. Hence, as the group is descended the enthalpy of solution (shown on the diagram as ΔH) becomes increasingly more negative. This means that the solubility of the Group 2 hydroxides increases as the group is descended. This is shown in the diagrams for magnesium hydroxide and barium hydroxide.

In the Group 2 sulphates the hydration enthalpy decreases at a greater rate than the lattice enthalpy decreases. This results in the enthalpy of solution becoming more positive as the group is descended. Hence the solubility of the Group 2 sulphates decreases as the group is descended.

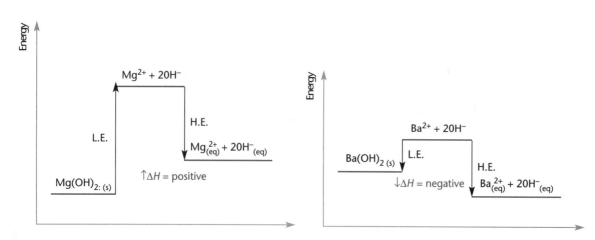

Energetics 2

Use your knowledge

1 Consider the diagram below:

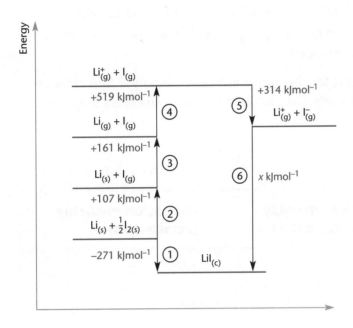

a) What name is given to the following enthalpy changes?

  i) 1 _____

  ii) 2 _____

  iii) 3 _____

  iv) 4 _____

  v) 5 _____

  vi) 6 _____

b) Using information in the diagram calculate the lattice enthalpy of lithium iodide.

c) Explain why the lattice enthalpy of lithium chloride is more exothermic than that of sodium iodide.

Consider what is happening to the lithium or iodine at each stage

Remember that the total energy changes on the left-hand side are equal to the total energy changes on the right-hand side

Consider percentage ionic character

Energetics 2

2 a) Write an equation for the lattice enthalpy of magnesium chloride.

b) Using the values below construct a Born–Haber cycle to calculate the lattice enthalpy of magnesium chloride

$\Delta H_{f(MgCl_2)} = -641$ kJ mol^{-1} $\Delta H_{at(Mg)} = +148$ kJ mol^{-1} First and second ionisation energy $+2189$ kJ mol^{-1} $\Delta H_{MgCl_2at(Cl_2)} = 121$ kJ mol^{-1} First electron affinity of chlorine $= -347$ kJ mol^{-1}

c) A data book gives the value for the lattice enthalpy of magnesium chloride as -2326 kJ mol^{-1}. Explain why the calculated value in b) differs from this value.

3 Consider the table of data below:

For solubility to occur the enthalpy of solution must be negative

	Lattice enthalpy (relative units)	Hydration enthalpy (relative units)
$MgSO_4$	100	150
$BaSO_4$	80	50

a) What is the trend in the solubility of the Group 2 sulphates as the group is descended?

b) Use the information in the table to justify this trend.

10 minutes

Test your knowledge

1. If the reactants in a reaction are thermodynamically stable with respect to the products, what is the sign of the enthalpy change?

2. What is entropy a measure of?

3. What is the sign of the entropy change when an ionic solid dissolves in water?

4. Write an expression for the Gibbs free energy change of a reaction in terms of the entropy change and the enthalpy change.

5. In terms of Gibbs free energy change, when is a reaction spontaneous?

 If you got them all right, skip to page 12

Thermodynamics

20 minutes

Improve your knowledge

1. Enthalpy, H^θ, is a measure of the internal energy of a system. The lower the enthalpy of one system compared to another, the more **thermodynamically stable** it is. Thus a **negative** enthalpy change, ΔH^θ, **favours** a given process (heat given out) and if ΔH^θ is **positive** the process is **not favoured** (heat taken in). Only ΔH^θ can be measured directly.

2. Entropy, S^θ, is a measure of the **disorder** of a system. The **greater the disorder**, i.e. S^θ, of a system, the more stable it is. For example, gases have more entropy than liquids, which have more entropy than solids. A system containing more moles than another also has more entropy. Entropy has units of $J\ K^{-1}\ mol^{-1}$.

3. A **positive** entropy change, ΔS^θ, **favours** a given process whereas a **negative** ΔS^θ is **unfavourable**. A positive ΔS^θ is responsible for why some **endothermic** reactions (ΔH^θ = +ve) *are* favourable. ΔS^θ can be calculated from the equation:

   $$\Delta S^\theta = \Sigma S^\theta(\text{products}) - \Sigma S^\theta(\text{reactants})$$

   Examples:

	$H_2O_{(s)}$	$\rightarrow$	$H_2O_{(l)}$	disorder increases, ΔS^θ = +ve
$S^\theta =$	48.0		69.9	ΔS^θ = +21.9 $J\ K^{-1}\ mol^{-1}$

	$2NO_{2(g)}$	$\rightarrow$	$N_2O_{4(g)}$	moles decrease, ΔHS^θ = –ve
$S^\theta =$	2×240		304	ΔS^θ = –176 $J\ K^{-1}\ mol^{-1}$

4. The overall thermodynamic **feasibility** of a reaction is governed by the change in **Gibbs free energy**, ΔG^θ. ΔG^θ is defined according to the equation:

   $$\Delta G^\theta = \Delta H^\theta - T\Delta S^\theta$$

   If ΔG^θ is ≤ 0 (i.e. negative) the reaction is feasible or **spontaneous**.
   The magnitude and sign of ΔG^θ depends upon ΔH^θ and $-T\Delta S^\theta$ as shown:

Thermodynamics

ΔH^θ	ΔS^θ	ΔG^θ	Spontaneity
+ve	−ve	+ve	No
-ve	+ve	−ve	Yes
+ve	+ve	+ve or −ve	Yes if $T\Delta S° > \Delta H°$
−ve	−ve	+ve or −ve	Yes if $\Delta H° > T\Delta S°$

Example

Calculate the Gibbs free energy change for the thermal decomposition of calcium carbonate at 25°C given that $\Delta H^\theta = +178$ kJ mol^{-1} and $\Delta S^\theta = +161$ J K^{-1} mol^{-1}.

$\Delta G^\theta = \Delta H^\theta - T\Delta S^\theta$
$\Delta G^\theta = +178 - 298 \times (+161/1000) = +130$ kJ mol^{-1}

Note that T must be converted to Kelvin (+273) and ΔS^θ must be converted to kJ (÷1000).

5 The temperature at which a reaction becomes spontaneous is when $\Delta G^\theta = 0$. This can be calculated from the equation in point 4.

If $\Delta G^\theta = 0$ then $T\Delta S^\theta = \Delta H^\theta$

therefore $T = \Delta H^\theta/\Delta S^\theta$

Example

Calculate the temperature at which the thermal decomposition of calcium carbonate becomes spontaneous, given that $\Delta H^\theta = +178$ kJ mol^{-1} and, $+ \Delta S^\theta = +161$ J K^{-1} mol^{-1}.

Spontaneity occurs when $\Delta G^\theta = 0$, i.e. $T\Delta S^\theta = \Delta H^\theta$.
Therefore, $T = \Delta H^\theta/\Delta S^\theta = +178 / (161/1000) = 1106$ K

ΔS^θ may be defined as $\Delta H^\theta/T$.

Thermodynamics

Use your knowledge

1

a) When solid potassium nitrate is dissolved in water the enthalpy change is slightly endothermic and yet the reaction is said to be spontaneous. Explain how this is possible.

b) Petrol is a mixture of liquid hydrocarbons and is widely used as a fuel. State and explain the sign of ΔH^θ, ΔS^θ and ΔG^θ.

c) Explain why liquid water only boils over 373 K, at atmospheric pressure.

Favourable ΔS°

Exothermic and an increase in moles

ΔG must be ≤ 0

2

At 298 K solid potassium nitrate dissociates when heated to produce solid potassium nitrite and gaseous oxygen according to the equation:

$$2KNO_{3(s)} \rightarrow 2KNO_{2(s)} + O_{2(g)}$$

The enthalpy change for this reaction = +124 kJ mol^{-1}.
The standard molar entropy, S^θ, for each species is given below.

	S^θ/ J K^{-1} mol^{-1}
$2KNO_{3(s)}$	133
$2KNO_{2(s)}$	152
O_2	205

a) Calculate the entropy change for the reaction.

b) Calculate the standard free energy change for the reaction.

c) Calculate the temperature at which this reaction becomes spontaneous.

Difference in total entropy of both sides

$\Delta G^\theta = \Delta H^\theta - T\Delta S^\theta$

Temperature when $\Delta G^\theta = 0$

Chemical Equilibria 2

15 minutes

Test your knowledge

1
a) What do you understand by the term 'dynamic equilibrium'?

b) Calculate K_c for the following reaction:

$$CH_3COOH_{(aq)} + C_2H_5OH_{(aq)} \rightleftharpoons CH_3COOC_2H_{5(aq)} + H_2O_{(l)}$$

given the equilibrium concentrates are 0.2, 0.2, 0.4 and 0.4 mol dm^{-3} respectively.

c) For the following equilibrium

$$N_{2(g)} + O_{2(g)} \rightleftharpoons 2NO_{(g)}$$

3.15 moles of each N_2 and O_2 were sealed in a 1 dm^3 container. At equilibrium 0.2 moles of NO were present. Calculate K_c.

2 For the following equilibrium:

$$N_{2(g)} + 3H_{2(g)} \rightleftharpoons 2NH_{3(g)} \quad \Delta H = -92 \text{ kJ mol}^{-1}$$

State the effect of:

a) An increase in the total pressure

b) An increase in nitrogen concentration.

c) A decrease in temperature.

d) The presence of a catalyst on the equilibrium position and the value of K_c.

3 Nitrogen monoxide dissociates into nitrogen and oxygen. 1 mole of NO was placed in a sealed container at 60 atmospheres pressure. At equilibrium the nitrogen monoxide was 20% dissociated. Calculate K_p.

Answers

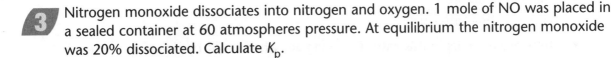

3 $K_p = 0.016$ (no units)

d) Catalysts do not alter K_c or move the equilibrium. A catalyst just reduces the time taken for equilibrium to be established.

c) The reaction favours the exothermic reaction, i.e. will move to the RHS, K_c will increase.

b) Equilibrium moves to the RHS, K_c remains unchanged.

2a) Equilibrium moves to the RHS (favouring the reaction that produces the fewer number of moles), K_c remains unchanged.

1a) A reversible reaction where the rate of the forward reaction is equal to the rate of the reverse reaction. b) 4 (no units) c) 4.3×10^{-3} (no units)

✓ **If you got them all right, skip to page 16**

13

Chemical Equilibria 2

20 minutes

Improve your knowledge

1 A reversible reaction is said to be at **(dynamic) equilibrium** when the rate of the forward reaction is equal to the rate of the reverse reaction. For the following reaction at equilibrium $aA + bB \rightleftharpoons cC + dD$ the ratio

$$\frac{[C]^c[D]^d}{[A]^a[B]^b}$$

Key points from
AS in a Week

Equilibria
pages 43–44

is constant. This constant is called the **equilibrium constant, K_c.** The value of K_c for a given reaction at a given temperature is always constant.

One mole each of ethanoic acid and ethanol were left to reach equilibrium. At equilibrium the amounts of ethanoic acid and ethanol were 0.33. **Calculate** the value of K_c.

The equation for this reaction is:

$$CH_3COOH_{(aq)} + C_2H_5OH_{(aq)} \rightleftharpoons CH_3COOC_2H_{5(aq)} + H_2O_{(l)}$$

- The first step is to determine the equilibrium concentrations of each substance. 0.67 (1 – 0.33) moles each of acid and alcohol reacted so 0.67 moles of the ester and water were formed.

- Thus $K_c = \dfrac{[0.67][0.67]}{[0.33][0.33]} = 4.1$

- There are no units in this case. The units are worked out by doing the same to the units as was done to the numerical values.

$$\frac{mol\ dm^{-3} \times mol\ dm^{-3}}{mol\ dm^{-3} \times mol\ dm^{-3}}$$

In this case the units completely cancel each other out.

2 **Le Chatelier's principle** states that when a change is imposed upon a system at equilibrium the system acts to oppose that change. Consider an increase in the **concentration** of ethanoic acid in the following equilibrium:

$$CH_3COOH_{(aq)} + C_2H_5OH_{(aq)} \rightleftharpoons CH_3COOC_2H_{5(aq)} + H_2O_{(l)}$$

The system wants to remove the extra acid and so the equilibrium position moves to the right. A new equilibrium will be established with the same value of K_c. For gaseous equilibrium (see below) the effect of **pressure** is the same as concentration.

The effect of **temperature** depends upon whether the reaction is exothermic or endothermic. If the forward reaction is exothermic (e.g. in the Haber process) then an increase in temperature favours the endothermic path (i.e. the reverse reaction). However, an increase in temperature increases the kinetic energy of the gas molecules, hence there will be more successful collisions. As a result a compromise temperature is used in the manufacture of ammonia.

The presence of a **catalyst** in equilibrium reactions (e.g. Fe in the Haber process) serves only to reduce the time taken for equilibrium to be reached. The catalyst does not increase the yield.

3 Equilibrium systems that involve gases can have a **gaseous equilibrium constant**, K_p. Instead of concentration terms being used, partial pressures are used. The partial pressure of a gas is equal to the mole fraction of the gas multiplied by the total pressure. Solids and liquids do not appear in K_p expressions.

In a 1 dm^3 container 1 mole of N_2O_4 forms an equilibrium mixture with NO_2. At equilibrium the N_2O_4 was found to be 60% dissociated. The total pressure was 1 atm. Calculate K_p.

$$N_2O_4 \rightleftharpoons 2NO_2$$

- The number of moles of N_2O_4 remaining at equilibrium is 0.4 (60% of 1 mole has dissociated). The number of moles of NO_2 is thus 1.2 (2 moles of NO_2 are formed for every 1 mole of N_2O_4 that dissociates). The total number of moles of gas present at equilibrium is 1.6 (1.2 + 0.4). The mole fraction of N_2O_4 is thus 0.4/1.6 and NO_2 1.2/1.6. The partial pressures are obtained by multiplying the mole fractions by the total pressure.

- Hence $K_p = \dfrac{P_{NO_2}^2}{P_{N_2O_4}} = \dfrac{0.75^2}{0.25} = 2.25$ atm   Units: (atm^2/atm = atm)

15 minutes

Use your knowledge

1   Consider the following equilibrium:

$$CH_3COOH + C_2H_5OH \rightleftharpoons CH_3COOC_2H_5 + H_2O \quad \Delta H = \text{negative}$$

30 g of ethanoic acid and 23 g of ethanol were mixed and allowed to reach equilibrium at 60°C. At equilibrium the number of moles of ester was 0.33.

a) Calculate K_c and state the units.

b) Explain the effect if increasing the temperature on:

   i) The equilibrium position
   ii) The value of K_c.

Calculate the number of moles of each reactant at the start

2   An important reaction in the contact process is the catalytic oxidation of sulphur dioxide using a vanadium (V) oxide catalyst:

$$2SO_{2(g)} + O_{2(g)} \rightleftharpoons 2SO_{3(g)} \quad \Delta H = -190 \text{ kJ mol}^{-1}$$

In an experiment to determine K_p sulphur dioxide and oxygen were mixed in the ratio (2 mol : 1 mol) and allowed to reach equilibrium in the presence of a catalyst, at a pressure of 5 atm. At equilibrium one third of the SO_2 had been converted to SO_3.

a) Calculate the value of K_p and state the units.

b) State and explain the effect of an increase in total pressure on the yield of sulphur trioxide.

c) State and explain the effect of a decrease in temperature on the value of K_p.

Consider the number of moles of gas on either side of the equilibrium

15 minutes

Test your knowledge

1 In each of the equilibrium systems below identify the acid, base, conjugate acid and conjugate base:

a) $CH_3COOH_{(aq)} + H_2O_{(l)} \rightleftharpoons CH_3COO^-_{(aq)} + H_3O^+_{(aq)}$

b) $H_2SO_{4(aq)} + HNO_{3(aq)} \rightleftharpoons HSO^-_{4(aq)} + H_2NO^+_{3(aq)}$

2 Calculate the pH of the following solutions:

a) 0.1 mol dm^{-3} HCl solution.

b) 0.3 mol dm^{-3} KOH solution.

3 Calculate the pH of the following weak acids:

a) 0.2 M ethanoic acid ($K_a = 1.8 \times 10^{-5}$)

b) 0.1 M benzoic acid ($K_a = 6.3 \times 10^{-5}$)

4 Calculate the pH of a buffer solution containing 0.2 mol dm^{-3} ethanoic acid and 0.4 mol dm^{-3} sodium ethanoate. (K_a ethanoic acid = 1.8×10^{-5})

5 a) Draw a curve of pH versus volume when 0.1 M sodium hydroxide is added to 25 cm^3 of 0.1 M ethanoic acid.

b) Suggest why methyl orange would not be a suitable indicator for this titration.

Answers

5a) See diagram page 20 b) pK_{In} needs to be in the range of the end-point.

4) 5.04

3a) 2.72 b) 2.6

2a) 1 b) 13.48

1a) A = CH_3COOH, B = H_2O, CA = H_3O^+, CB = CH_3COO^-

b) Acid = H_2SO_4, base = HNO_3, conjugate acid = $H_2NO_3^+$, conjugate base = HSO_4^-

 If you got them all right, skip to page 22

Acids and Bases

20 minutes

Improve your knowledge

 Brønsted-Lowry theory describes **acids** as **proton donors**. Consider an aqueous solution of hydrochloric acid:

$$HCl_{(aq)} + H_2O_{(l)} \rightarrow H_3O^+_{(aq)} + Cl^-_{(aq)}$$

HCl is a Brønsted-Lowry acid because it donates a proton to H_2O. The Cl^- anion is the **conjugate base** of HCl. An acid and its conjugate base are related by one proton.

Key points from
AS in a Week
Equilibria
            page 43–44

2 **pH** is defined as the negative logarithm to the base 10 of the concentration of oxonium (H_3O^+) ions. This is represented symbolically in one of the following terms:

$$pH = -\log_{10}[H_3O^+] \text{ or } pH = -lg[H_3O^+] \text{ or } pH = -\log_{10}[H^+]$$

An acid or base is described as strong if it completely dissociates in aqueous solution.

$$HNO_{3(aq)} + H_2O_{(l)} \rightarrow H_3O^+_{(aq)} + NO^-_{3(aq)}$$

The pH of a strong monoprotic acid is calculated by using the concentration of the acid as the concentration of the oxonium ions.

Calculate the pH of 0.2 M hydrochloric acid.

$$pH = -\log_{10}[H_3O^+]$$
$$pH = -\log_{10}[0.2]$$
$$pH = 0.70$$

N.B. for diprotic acids the concentration of oxonium ions will be twice the acid concentration.

The pH of strong bases can be calculated by substituting the value of pOH into the equation:

$$pH + pOH = 14 \quad pOH = -\log_{10}[OH^-].$$

Acids and Bases

 Weak acids only partially dissociate in aqueous solution, hence an equilibrium system is set up.

Consider the weak acid HA. It will dissociate as follows:

$$HA_{(aq)} + H_2O_{(l)} \rightleftharpoons H_3O^+_{(aq)} + A^-_{(aq)}$$

The acid dissociation constant (K_a) is defined as:

$$K_a = \frac{[H_3O+] \, [A^-]}{[HA]}$$

Knowing the K_a value for an acid and its concentration allows the pH to be found.

Calculate the pH of 0.25 M ethanoic acid ($K_a = 1.8 \times 10^{-5}$)

$$1.8 \times 10^{-5} = \frac{[CH_3COO^-] \, [H_3O^+]}{[CH_3COOH]}$$

Assumptions:

- $[CH_3COO^-] = [H_3O^+]$
- $[CH_3COOH]$ remains constant

Incorporating these assumptions into the K_a expression:

$$1.8 \times 10^{-5} = \frac{[H_3O^+]^2}{0.25}$$
$$[H_3O^+] = 2.12 \times 10^{-3}$$
$$pH = 2.67$$

 A buffer solution is one which resists changes in pH caused by addition of small amounts of acid, base or upon dilution. Buffer solutions are solutions of weak acids (or bases) and salts of the corresponding acid (or base). A mixture of ethanoic acid and sodium ethanoate is a buffer, as is a mixture of ammonia and ammonium chloride.

The pH of a buffer solution is calculated using the following equation:

$$pH = pK_a + \log_{10}\left(\frac{[salt]}{[acid]}\right)$$

N.B. The pK_a value is $-\log_{10}K_a$

Buffers work because the weak acid (HA) only partially dissociates, whilst the salt (MA) completely dissociates.

$$HA_{(aq)} + H_2O_{(l)} \rightleftharpoons H_3O^+_{(aq)} + A^-_{(aq)}$$

$$MA_{(s)} + aq \rightarrow M^+_{(aq)} + A^-_{(aq)}$$

Hence the mixture contains a high concentration of acid (the largely undissociated HA) and a high concentration of base (the conjugate base of HA). Acid (H$^+$) added is removed by combination with the A$^-$ to form undissociated HA (according to Le Chatelier's principle). Base (OH$^-$) added is removed by combination with the H$_3$O$^+$ ions from the dissociated acid. This then causes more of the acid to dissociate (by Le Chatelier's principle).

The graph shows the changes in pH that occur when **titrations** are carried out using strong and weak acids and bases.

The **indicator** used for each titration depends upon the pK_{In} value of the indicator and the pH of the end point. Consider the indicators and their pK_{In} values in the table below.

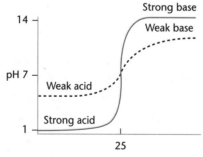

Volume of 0.1 M base added to
25 cm^3 0.1 M acid

Consider the above titration curve for a strong acid – weak base titration. A suitable indicator would be either methyl orange or bromothymol blue, because their pK_{In} values lie on the vertical part of the titration curve. Phenolphthalein would be unsuitable.

Indicator	pK_{In}
Methyl orange	4.0
Bromothymol blue	7.0
Phenolphthalein	9.7

Make sure you know the following:

the Brønsted-Lowry definition of an acid / how to identify conjugate acid–base pairs / how to calculate pH of strong and weak acids / how to calculate the pH of buffer solutions / how to plot titration curves / how to select an indicator

Acids and Bases

15 minutes

Use your knowledge

1. This question considers two acids, hydrochloric and ethanoic, which have different properties.

   a) Define pH.

   Learn this!

   b) Calculate the pH of a 0.25 M solution of hydrochloric acid.

   Strong acids completely dissociate

   c) Calculate the pH of this acid if 10 cm³ of 0.1 M sodium hydroxide is added to 50 cm³ of this acid.

   Calculate the concentration of the remaining acid

   Hydrochloric acid is described as a strong acid whilst ethanoic acid is described as a weak acid.

   d) What do you understand by the term 'weak acid'?

   Consider dissociation

   e) Calculate the pH of a 0.42 M solution of propanoic acid pKa = 4.9.

   Remember the two approximations

   f) Calculate the pH of a solution containing 0.2 M propanoic acid and 0.6 M sodium propanoate.

   What kind of solution is this?

2. A student was titrating a solution of 0.1 M ethanoic acid with a strong base of similar concentration.

   a) Sketch a titration curve for this titration.

   Learn these curves

   b) Suggest a base that could have been used to produce a curve such as the one above.

   Look at the final pH

   c) From the table of pK_{In} values on page 21 suggest, giving a suitable reason, a suitable indicator for this titration.

   Consider the pK_{In} value

Test your knowledge

1 a) Write an expression for the rate equation for a reaction that is first order with respect to reactant A and zero order with respect to reactant B.
b) What is the overall order of the reaction?

2 Outline methods of measuring the rate of each of the following reactions:
a) $CACO_{3(s)} + 2HCl_{(aq)} \rightarrow CaCl_{2(aq)} + H_2O_{(l)} + CO_{2(g)}$
b) $Cl_{2(g)} + 2I^-_{(aq)} \rightarrow 2Cl^-_{(aq)} + I_{2(g)}$

3 Experimental data obtained for the kinetics of the reaction $X + 2Y \rightarrow XY_2$ are summarised below:

Experiment	Initial [X]	Initial [Y]	Initial rate/mol dm^{-3} s^{-1}
1	0.25	0.25	0.2
2	0.25	0.125	0.1
3	0.35	0.1	0.112
4	0.175	0.1	0.056

a) Deduce the order of reaction with respect to X.
b) Deduce the order of reaction with respect to Y.
c) Write an expression for the rate equation.
d) Calculate the value of k.
e) State the units of k.

4 Many reactions take place in many steps. What name is given to the slowest step?

Answers

4 The rate-determining step.
3a) 1 b) 1 c) rate = $k[X]^1[Y]^1$ d) 3.2 e) mol^{-1} dm^3 s^{-1}
2a) Measuring the volume of carbon dioxide produced. b) Titration of the iodine (using sodium thiosulphate).
1a) rate = $k[A]^1[B]^0$ (or rate = $k[A]^1$) b) 1

 If you got them all right, skip to page 27

Improve your knowledge

1 For the reaction A + B → Products, the **rate equation** is

$$\text{rate} = k[A]^x[B]^y$$

Key points from
AS in a Week

Kinetics 1
        page 38–40

where k is the rate constant and is simply a constant of proportionality. The square brackets represent the concentration, and x and y are the individual **orders of reaction** of A and B respectively. The order of reaction is the numerical power to which each of the concentrations of reactants is raised in the rate equation. The value of $x + y$ is the overall order of reaction and rarely exceeds 2. Both the rate constant and the orders of reaction can only be determined experimentally.

2 To determine the above values, methods of **measuring the rate of a reaction** need to be established in order to derive the rate equation. Methods include:

- Collecting gas evolved
- Colourimetry
- Titration.

Often it is necessary to **quench** a reaction (i.e. rapidly stop it) before carrying out a titration. Quenching allows the concentration of one or more reactants to be determined without the reaction continuing.

3 The rate of equation for the reaction between A and B can be deduced from the data below by considering the effect on the rate of **varying the concentration** of either A or B.

Experiment	Initial [A]	Initial [B]	Initial rate/mol dm^{-3} s^{-1}
1	0.1	0.1	0.001
2	0.1	0.2	0.004
3	0.1	0.3	0.009
4	0.2	0.1	0.001
5	0.3	0.1	0.001

If the concentration of one reactant doubles and the rate doubles, the order of reaction with respect to that reactant is **1**. If the concentration doubles and the rate quadruples then the order is **2**. If the rate is independent of the concentration then the order is **0**.

Consider experiments 1 and 2. The concentration of A remains constant but the concentration of B doubles. The effect of this on the rate of the reaction is that the rate quadruples, hence the order of reaction with respect to B is 2.

Similarly, in experiments 1 and 4 (or 5), the concentration of B remains constant yet the concentration of A doubles (or triples). However this has no effect on the rate and hence the order of reaction with respect to A is 0.

Thus the rate equation becomes rate = $k[A]^0[B]^2$. The numerical value of k can be found by substituting the values for any experiment (they will all give the same value). Using experiment 3:

$$0.009 = k\,[0.1]^0[0.3]^2$$
$$0.009 = k\,0.09$$
$$k = 0.1$$

The **units of k** can be worked out in a similar way (by using the units rather than the numerical values) or they can be learnt.

Order of reaction	Units of k
0	$mol\ dm^{-3}\ s^{-1}$
1	s^{-1}
2	$mol^{-1}\ dm^3\ s^{-1}$

Many reactions take place in several steps. Each step will occur at a different rate. The step that has the slowest rate is known as the **rate-determining step**. For a reaction that occurs in more than one step the rate equation can be determined and information about the mechanism of the reaction can be deduced.

Consider the following reaction:

$$CH_3CSNH_2 + 2OH^- \rightarrow CH_3CO_2^- + HS^- + NH_3$$

The reaction is known to occur in two steps and the rate equation for this reaction shows that the reaction is first order with respect to both thioethanamide (CH_3CSNH_2) and hydroxide ions.

As the reaction is first order with respect to both reactants, the first stage of the reaction is therefore known to contain one of each species. In fact the rate-determining step is:

$$CH_3CSNH_2 + 2OH^- \rightarrow CH_3CONH_2 + HS^-$$

Given that the overall reaction is known and the rate-determining step is known, the second step of the reaction can be proposed:

$$CH_3CONH_2 + OH^- \rightarrow CH_3CO_2^- + NH_3$$

By adding up these two steps the overall equation is obtained.

Use your knowledge

 The iodination of propanone can be respresented as:

$$CH_3COCH_3 + I_2 + H^+ \rightarrow CH_2ICOCH_3 + I^- + 2H^+ \text{ (H}^+ \text{ acts as a catalyst)}.$$

Initial rate data for this reaction is shown in the table below:

Experiment	$[I_2]$	$[CH_3COCH_3]$	$[HCl]$	rate
1	0.001	0.5	1.25	1.1×10^{-5}
2	0.002	0.5	1.25	1.1×10^{-5}
3	0.002	1.0	1.25	2.2×10^{-5}
4	0.002	1.0	2.5	4.4×10^{-5}

a) Deduce the order of reaction for each of the reactants.

b) Calculate k and state its units.

c) State how the concentration of iodine can be measured over the course of this reaction.

 The following reaction is catalysed by Fe^{2+} ions.

$$S_2O_8^{2-}{}_{(aq)} + 2I^-_{(aq)} \rightarrow 2SO_4^{2-}{}_{(aq)} + I_{2(l)}$$

Kinetic studies show that the rate-determining step is first order with respect to $S_2O_8^{2-}$ ions and zero order with respect to I^- ions. The first step of the reaction is:

$$S_2O_8^{2-}{}_{(aq)} + 2Fe^{2+}{}_{(aq)} \rightarrow 2SO_4^{2-}{}_{(aq)} + 2Fe^{3+}{}_{(aq)}$$

Write an equation for the second step of this reaction.

The second step added to the first step must equal the overall equation

Redox Equilibria

Test your knowledge

1 For the electrode potential of a half-cell to be measured, it needs to be connected to a standard hydrogen electrode. For Cu^{2+}/Cu the $E^{\ominus}$ value is $+ 0.34$ V.

a) Draw and label a copper half-cell.

b) Draw and label a standard hydrogen electrode.

c) State and explain how the two half-cells would be connected in order for the cell e.m.f. to be calculated.

d) What piece of apparatus is used to measure the cell e.m.f.?

2 a) Write the cell equation for the galvanic cell in Question 1.

b) Calculate the cell e.m.f. for the following galvanic cell.

$$Zn_{(s)} \mid Zn^{2+}_{(aq)} \parallel Fe^{2+}_{(aq)} \mid Fe_{(s)}$$

$$E^{\ominus} (Fe^{2+}_{(aq)}/Fe_{(s)}) = -0.44 \text{ V} \qquad E^{\ominus} (Zn^{2+}_{(aq)}/Zn_{(s)}) = -0.77 \text{ V}$$

c) Write the cell equation for the galvanic cell consisting of the copper and iron half-cells.

d) Calculate the e.m.f. for this cell.

3 Predict the overall reaction that occurs when the following two half-cells are combined:

$$Zn^{2+} + 2e^- \rightarrow Zn \qquad E^{\ominus} = - 0.76 \text{ V}$$
$$Fe^{2+} + 2e^- \rightarrow Fe \qquad E^{\ominus} = + 0.80V$$

4 Using the information below, explain why lead should not be used as a coating in order to prevent iron rusting.

$$Pb^{2+} + 2e^- \rightarrow Pb \qquad E^{\Theta} = -0.13 \text{ V}$$
$$Fe^{2+} + 2e^- \rightarrow Fe \qquad E^{\Theta} = -0.44 \text{ V}$$

5 What are the principal differences between a galvanic cell and a storage cell?

Answers

1a) A copper rod (any size) immersed in a 1M solution of any copper (II) salt, e.g. copper sulphate. Temperature of the solution should be 298 K.
b) See diagram page 30.
c) A salt bridge (a piece of filter paper soaked in saturated potassium nitrate or potassium chloride) completes the circuit, allows flow of ions (to even charge) but no chemical reaction occurs.
d) A potentiometer (or high resistance voltmeter).
2a) $Pt \mid H_{2(g)}, H^+_{(aq)} \parallel Cu^{2+}_{(aq)} \mid Cu_{(s)}$
b) +33 V
c) $Fe_{(s)} \mid Fe^{2+}_{(aq)} \parallel Cu^{2+}_{(aq)} \mid Cu_{(s)}$
d) +0.78 V
3) $Zn + 2Ag^+ \rightarrow Zn^{2+} + 2Ag$
4) The overall reaction is $Fe + Pb^{2+} \rightarrow Fe^{2+} + Pb$, i.e. rusting occurs.
5) Storage cells must produce a current and be transportable.

 If you got them all right, skip to page 34

20 minutes

Improve your knowledge

1 A metal immersed in a solution of its own ions sets up the equilibrium $M^{n+}_{(aq)} + ne^- \rightleftharpoons M_{(s)}$. Depending on which process predominates, the metal either gains a positive or negative charge. Thus there is a **potential difference** between the **metal** and the **solution**. This system is known as a **half-cell**. The potential difference of a half-cell depends upon the metal used, the temperature and the concentration of the salt solution. The temperature and concentration are standardised (298 K and 1 mol dm^{-3}) to allow standard electrode potentials ($E^{\ominus}$) to be measured. The electrode potential of a half-cell cannot be measured directly, but the potential difference between two electrodes can be measured. The **standard hydrogen electrode** is taken as a reference electrode. It is arbitrarily assigned an $E^{\ominus}$ value of 0.0 V.

Key points from AS in a Week

Equilibria
page 43–44

Introduction to oxidation and reduction
pages 29–31

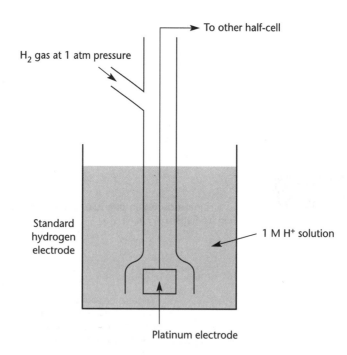

To other half-cell

H$_2$ gas at 1 atm pressure

Standard hydrogen electrode

1 M H$^+$ solution

Platinum electrode

Redox Equilibria

The combination of two half-cells gives a voltaic (or galvanic) cell. The cells need to be connected via a salt bridge. This is often a piece of filter paper soaked in saturated potassium nitrate solution. The salt bridge allows the circuit to be completed; it allows migration of ions within the salt bridge (to neutralise the charge) but no chemical reaction.

The maximum potential difference of the cell is known as the e.m.f. (electromotive force). It occurs when no current flows, so a conventional voltmeter cannot be used to determine the e.m.f. A potentiometer (or high resistance voltmeter) is used instead.

2 In a **galvanic cell** one electrode experiences **reduction** and the other **oxidation**. It is possible to predict the overall cell reaction (as a combination of **redox** processes) using tables of $E^{\ominus}$ values (from data books).

First the cell equation should be written. A galvanic cell containing zinc and copper half-cells would have a cell equation written as:

$$Zn_{(s)} \mid Zn^{2+}_{(aq)} \parallel Cu^{2+}_{(aq)} \mid Cu_{(s)}$$

Vertical lines represent phase boundaries, double vertical lines represent the salt bridge. The solutions are written either side of the salt bridge. The more positive electrode is written on the right-hand side (Cu/Cu^{2+} $E^{\ominus} = 0.34$ V, $E^{\ominus} = -0.76$ V)

The cell e.m.f. is calculated by:

$$E^{\ominus}_{Cell} = E^{\ominus}_{Right\text{-}hand\ electrode} - E^{\ominus}_{Left\text{-}hand\ electrode}$$
$$= 0.34 - (-0.76)$$
$$= +1.10 \text{ V}$$

The copper electrode is the more positive so electrons will flow from the zinc to the copper. Hence the reaction at the zinc electrode is $Zn \rightarrow Zn^{2+} + 2e^-$ (i.e. producing electrons) and the reaction at the copper electrode is $Cu^{2+} + 2e^- \rightarrow Cu$. Adding these two half-equations gives an overall equation of $Zn + Cu^{2+} \rightarrow Zn^{2+} + Cu$.

The cell e.m.f. assumes standard conditions within each half-cell. The effect of non-standard temperatures and concentrations on the $E^{\ominus}$ value of each half-cell can be predicted by Le Chatelier's principle on the $M^{n+}_{(aq)} + ne^- \rightleftharpoons M_{(s)}$ equilibrium (the forward reaction is the exothermic pathway). The effect on each half-cell can then be used to predict the effect on the cell e.m.f.

Redox Equilibria

3 The likely direction of spontaneous change of redox reactions can be predicted using standard electrode potential data.

Consider the two half-equations below:

$$Sn^{4+} + 2e^- \rightarrow Sn^{2+} \qquad E^\ominus = +0.15 \text{ V}$$
$$Fe^{3+} + e^- \rightarrow Fe^{2+} \qquad E^\ominus = +0.77 \text{ V}$$

It is apparent that Fe^{3+} has a greater attraction for electrons than Sn^{4+}. Therefore if Fe^{3+} is to be reduced, then something must be oxidised. Clearly, the Sn^{2+} is the only species that can be oxidised. As a result the two half-equations occurring in this redox system are:

$$Fe^{3+} + e^- \rightarrow Fe^{2+}$$
$$Sn^{2+} \rightarrow Sn^{4+} + 2e^-$$

Combination of these half-equations shows that the overall reaction is:

$$Sn^{2+} + 2Fe^{3+} \rightarrow Sn^{4+} + 2Fe^{2+}$$

The cell is therefore written as

$$Sn^{2+}_{(aq)} \mid Sn^{4+}_{(aq)} \parallel Fe^{3+}_{(aq)} \mid Fe^{2+}_{(aq)}$$

$$E_{cell} = E_{RHS} - E_{LHS} = 0.77 - 0.15 = 0.62 \text{ V}$$

Note that this predicted reaction only works under standard conditions, i.e. a temperature of 298 K and all solutions are 1 mol dm^{-3}. It is generally considered that cell reactions will proceed at a reasonable rate and go to completion if E_{cell} exceeds 0.4 V.

4 Knowledge of redox equilibria is used to understand rusting and storage cells.

Corrosion is the conversion of a metal (usually iron) in its normal working environment to its ions. Corrosion of all metals is an **oxidation reaction**.

The anodic reaction occurring is: $Fe_{(s)} \rightarrow Fe^{2+}_{(aq)} + 2e^-$

The cathodic reaction occurring is : $O_{2(aq)} + 2H_2O_{(l)} + 4e^- \rightarrow 4OH^-_{(aq)}$

If these reactions occur in the close proximity then $Fe(OH)_2$ is precipitated. Oxygen in the air oxidises this to hydrated iron(III) oxide(rust), $Fe_2O_3 \cdot xH_2O$.

Redox Equilibria

There are many methods of preventing rusting. One is to coat iron with a metal which has a more negative electrode potential. Zinc is often used to coat iron (galvanising). The $E^{\ominus}$ values are shown below:

$$Fe^{2+}_{(aq)} + 2e^- \rightarrow Fe \qquad E^{\ominus} = -0.44V$$

$$Zn^{2+}_{(aq)} + 2e^- \rightarrow Zn \qquad E^{\ominus} = -0.76V$$

Combination of these two half-equations shows that the overall reaction is:

$$Zn + Fe^{2+} \rightarrow Fe + Zn^{2+}$$

i.e. the zinc is oxidised/corrodes in preference to the iron.

Storage cells differ from galvanic cells in that they must produce a current. In addition, they must be transportable without leakage. The negative pole of a storage cell produces electrons while the positive pole accepts them.

The lead–acid cell is found in all car engines (as the car battery). The anode is made from lead and the cathode from lead(II) oxide. The electrolyte in which the electrodes are immersed is dilute sulphuric acid.

Anode equation: $\quad Pb + SO_4^{2-} \rightarrow PbSO_4 + 2e^- \qquad\qquad\qquad\quad E^{\ominus} = -0.36V$

Cathode equation $\quad PbO_2 + SO_4^{2-} + 4H^+ + 2e^- \rightarrow PbSO_4 + 2H_2O \quad E^{\ominus} = +1.69V$

These reactions are known as discharge reactions. Charging (by the vehicle's alternator) reverses the above reactions.

Redox Equilibria

Use your knowledge

1 The diagram represents a galvanic cell consisting of nickel and silver half-cells. The $E^\ominus$ value for the Ni/Ni^{2+} electrode is -0.25 V.

a) Identify A and B and state the concentration of the solutions used.

b) Write the cell equation for this galvanic cell.

c) The reading on A is 1.05 V. Calculate $E^\ominus$ for the silver half-cell.

d) What is the overall cell reaction?

e) What will be the effect of diluting the silver nitrate solution on:

  i) The $E^\ominus$ value of the silver half-cell
  ii) The overall cell e.m.f.?

2 The standard electrode potentials of the reactions involved in the first stage of the rusting of iron are:

$$Fe^{2+}_{(aq)} + 2e^- \rightarrow Fe \qquad\qquad E^\ominus = -0.44 \text{ V}$$

$$O_{2(aq)} + 2H_2O_{(l)} + 4e^- \rightarrow 4OH^-_{(aq)} \qquad E^\ominus = +0.44 \text{ V}$$

a) Write the overall equation for the first stage of the rusting of iron.

b) Explain how a piece of zinc metal attached to a sheet of iron prevents it from rusting.

Test your knowledge

1 a) Write balanced symbol equations for each of the following reactions:

   i) sodium + chlorine
   ii) magnesium + chlorine
   iii) aluminium + water
   iv) phosphorus + oxygen
   v) chlorine + water.

b) Complete the table below on the acid–base nature of some Period 3 oxides/hydroxides:

Substance	Formula	Bonding	Acid–base nature
Sodium oxide		Ionic	
Magnesium hydroxide	$Mg(OH)_2$		
Aluminium hydroxide			Amphoteric
Silicon dioxide	SiO_2		
Sulphur trioxide		Covalent	
Chlorine dioxide			Acidic

2 a) What is the trend in metallic character in the Group 4 elements?

b) How does the stability of the +4 oxidation state in the compounds of Group 4 elements change as the group descended?

3 a) State the acid–base nature of carbon dioxide.

b) Write equations to represent the amphoteric nature of lead(II) oxide (PbO).

 4 a) What is the shape of a carbon tetrachloride molecule?

b) Write an equation for the reaction of silicon tetrachloride with water.

Answers

1) See tables on pages 37 and 38.
2a) Metallic character increases down the group.
b) +4 becomes less stable down the group.
3a) Carbon dioxide is (weakly) acidic.
b) See Section 3, page 40.
4a) Tetrahedral
b) $SiCl_4 + 4H_2O \rightarrow SiO_2 + 4HCl$

 ✔ **If you got them all right, skip to page 41**

Periodic Table 2

10 minutes

Improve your knowledge

1 The reactions of the Period 3 elements (sodium to chlorine) with oxygen, chlorine and water as well as a consideration of the formulae and acid–base nature of some of their compounds can illustrate the variation in properties across a period.

Key points from
AS in a Week

Periodic table 1
    page 25–26

Groups 1 + 2
    pages 47–48

Element	Reaction with oxygen	Reaction with chlorine	Reaction with water
Na	$4Na_{(s)} + 2O_{2(g)} \rightarrow Na_2O_{(s)}$	$2Na_{(s)} + Cl_{2(g)} \rightarrow 2NaCl_{(s)}$	$2Na_{(s)} + 2H_2O \rightarrow 2NaOH_{(aq)} + H_{2(g)}$
Mg	$2Mg_{(s)} + O_{2(g)} \rightarrow 2MgO_{(s)}$	$Mg_{(s)} + Cl_{2(g)} \rightarrow MgCl_{2(s)}$	$Mg_{(s)} + H2O_{(l)} \rightarrow MgO_{(s)} + H_{2(g)}$
Al	$4Al_{(s)} + 3O_{2(g)} \rightarrow Al_2O_{3(s)}$	$2Al_{(s)} + 3Cl_{2(g)} \rightarrow 2AlCl_{3(s)}$	$2Al_{(s)} + 3H_2O_{(g)} \rightarrow Al_2O_{3(s)} + 3H_{2(g)}$
Si	$Si_{(s)} + 2O_{2(g)} \rightarrow SiO_{2(s)}$	$Si_{(s)} + 2Cl_{2(g)} \rightarrow SiCl_{4(l)}$	No reaction
P	$4P_{(s)} + 3O_{2(g)} \rightarrow P_4O_{6(s)}$ $4P_{(s)} + 5O_{2(g)} \rightarrow P_4O_{10(s)}$	$4P_{(s)} + 3Cl_{2(g)} \rightarrow PCl_{3(l)}$ $2P_{(s)} + 5Cl_{2(g)} \rightarrow 2PCl_{5(s)}$	No reaction
S	$S_{(s)} + O_{2(g)} \rightarrow SO_{2(g)}$ $2S_{(s)} + 3O_{2(g)} \rightarrow 2SO_{3(g)}$	$S_{(s)} + Cl_{2(g)} \rightarrow SCl_{2(g)}$ $2S_{(s)} + Cl_{2(g)} \rightarrow S_2Cl_{2(l)}$	No reaction
Cl	$2Cl_{2(g)} + O_{2(g)} \rightarrow 2Cl_2O_{(g)}$ $2Cl_{2(g)} + 7O_{2(g)} \rightarrow 2Cl_2O_{7(g)}$	–	$Cl_{2(g)} + H_2O_{(l)} \rightarrow HOCl_{(aq)} + HCl_{(aq)}$

Within the Period 3 oxides it is interesting to note that the ratio of oxygen atoms increases by $1/2$ each time as the period is ascended (considering the higher oxide where more than one stable oxide is formed).

The acid–base nature of the Period 3 metal oxides and hydroxides and non-metal oxides is shown in the table below.

Formula of substance	Acid–base nature	Bonding
Na_2O	Basic	Ionic
NaOH	Basic	Ionic
MgO	Weakly basic	Ionic (with a degree of covalent character)
$Mg(OH)_2$	Weakly basic	Ionic (with a degree of covalent character)
Al_2O_3	Amphoteric	Mixture of both ionic and covalent bonding
$Al(OH)_3$	Amphoteric	Mixture of both ionic and covalent bonding
SiO_2	Weakly acidic	Giant covalent
P_4O_{10}	Acidic	Covalent
SO_2	Acidic	Covalent
SO_3	Acidic	Covalent
Cl_2O	Acidic	Covalent

Note the relationship between the acid–base nature of the oxides and their bonding.

Periodic Table 2

The table below shows the formulae and properties of the Period 3 chlorides.

Formula	Bonding	Reaction with water	Equation for reaction with water
NaCl	Ionic	Dissolves	$NaCl + aq \rightarrow Na^+(aq) + Cl^-(aq)$
$MgCl_2$	Ionic	Soluble	$MgCl_2 + aq \rightarrow Mg^{2+}(aq) + 2Cl^-(aq)$
$AlCl_3$	Covalent	Reversible hydrolysis	$AlCl_3 + 3H_2O \rightarrow Al(OH)_3 + 3HCl$
$SiCl_4$	Covalent (giant)	Hydrolysis	$SiCl_4 + 2H_2O \rightarrow SiO_2 + 4HCl$
PCl_3 PCl_5	Covalent	Hydrolysis	$PCl_3 + 3H_2O \rightarrow H_3PO_3 + 3HCl$ $PCl_5 + 4H_2O \rightarrow H_3PO_4 + 4HCl$
SCl_2	Covalent	Hydrolysis	$2SCl_2 + 2H_2O \rightarrow SO_2 + S + 4HCl$

Note the increasing susceptibility of the chlorides to hydrolysis as the period is ascended.

 Group 4 elements show the following trends.

As Group 4 is descended the **increase in metallic character** is particularly noticeable. This is because with increasing atomic number there is an increasing atomic radius and hence a decreasing attraction for the outer electrons. This causes delocalisation, which is characteristic of metals.

The +2 oxidation state becomes **increasingly more stable** as the group is descended. This can be shown by considering lead and tin. Sn^{2+} is a reducing agent (itself being oxidised to +4) showing that the +4 oxidation state for tin is more stable. Conversely, the +4 oxidation state of lead is oxidizing (itself being reduced to +2).

The electronic configuration of tin is $5s^2\ 5p^2$. An oxidation state of +2 is obtained by losing the 2p electrons. An oxidation state of +4 is obtained by promoting one of the s electrons into the vacant p orbital, forming 4 sp^3 hybrid orbitals. The energy required to promote this electron is compensated for by the energy released upon forming two extra covalent bonds. Lead has the electronic configuration $6s^2 6p^2$. The promotion energy required to achieve an oxidation state of +4 is not compensated for by the formation of two extra bonds. This is because lead is much bigger than tin and so forms weaker bonds. This is called the 'inert pair' effect.

 As Group 4 is descended there is a clear pattern in the acid–base nature of the oxides:

Carbon forms two oxides, CO and CO_2. Carbon dioxide is weakly acidic, forming carbonic acid upon addition to water.

$$CO_{2(g)} + H_2O_{(l)} \rightleftharpoons H_2CO_{2(aq)}$$

Silicon dioxide forms a giant molecular compound which is weakly acidic:

$$SiO_{2(s)} + 2NaOH_{(aq)} \rightarrow Na_2SiO_{3(aq)} + H_2O_{(l)}$$

Lead forms three oxides, PbO, PbO_2 and Pb_3O_4. Each oxide exhibits amphoteric nature:

$$PbO + 2H^+ \rightarrow Pb^{2+} + H_2O$$
$$PbO + OH^- + H_2O \rightarrow Pb(OH)$$

$$PbO_2 + 4H^+_{(conc)} \rightarrow Pb^{4+} + 2H_2O$$
$$PbO_2 + 2OH^- \rightarrow PbO_3^{2-} + H_2O$$

PbO_3O_4 behaves as a mixture of PbO and PbO_2.

 Carbon tetrachloride (CCl_4) and silicon tetrachloride ($SiCl_4$) are both tetrahedral molecules. This is because they each have four bond pairs of electrons (and no lone pairs) surrounding the central atom. However, they have different reactions with water.

Silicon tetrachloride is easily hydrolysed because the low-level d sub-shell (3d) is available to accept electrons. Carbon tetrachloride is not hydrolysed due to the lack of a 2d sub-shell.

$$SiCl_4 + 4H_2O \rightarrow SiO_2 + 4HCl$$

10 minutes

 Use your knowledge

1 This question concerns Period 3 of the periodic table (sodium to chlorine).

a) Write equations for the reactions of sodium and sulphur with oxygen.

b) What is the acid–base nature of each of these oxides?

c) What is the connection between the acid–base nature of the oxides and the bonding present in each oxide?

d) Write equations for the reactions of magnesium chloride and phosphorus chloride with water.

e) What is the connection between the bonding present in each oxide and the susceptibility of each chloride to hydrolysis?

f) Using your answer to part e), explain the reaction of aluminium chloride with water.

Learn the information in the tables on pages 37–39.

2 This question concerns the trends in the Group 4 elements.

a) Explain why the metallic character of lead is greater than that of carbon.

b) What is the acid-base nature of the oxides of silicon and lead?

c) Use equations to justify your above answer.

d) Explain why tin (II) compounds are reducing agents but lead (IV) compounds are oxidising agents.

Metals have delocalised electrons

Learn this!

Learn these!

Inert pair effect

30 minutes

Test your knowledge

1 Give the outer-shell electronic structures of Cr and Fe^{2+}.

2 What feature of NH_3 makes it able to act as ligand and what feature of transition metals enables them to form complexes with NH_3?

3 When cobalt chloride dissolves in water it produces a pink solution. Give the name, formula and shape of the species responsible for this colour.

4 When aqueous NaOH is added to a solution of Al^{3+} ions a white precipitate is formed which dissolves in excess NaOH. Give the formula of the two Al species formed.

5 Give an equation to show why a solution of chromium sulphate is acidic.

6 When excess conc. HCl is added to a solution of copper sulphate a yellow solution results. Give the formula and shape of the copper species formed and the type of reaction.

7 Give three factors which affect the colour of a transition metal complex.

8 Give the oxidation states of manganese in the following species. MnO_4^-, MnO_2, $[Mn(H_2O)_4(OH)_2]$.

9 Give the catalyst in the Haber process. What type of catalyst is this?

Answers

9 Fe / heterogeneous
8 +7 / +4 / +2
7 Any of: metal/oxidation state/ligand type/co-ordination number.
6 $[CuCl_4]^{2-}$ / tetrahedral/ligand substitution
5 $[Cr(H_2O)_6]^{3+} + H_2O \rightarrow [Cr(H_2O)_5(OH)]^{2+} + H_3O^+$
4 $[Al(H_2O)_3(OH)_3]$ / $[Al(H_2O)_2(OH)_4]^-$
3 Hexaaquacobalt (II) ion/$[Co(H_2O)_6]^{2+}$/octahedral
2 Lone pair/vacant d orbitals.
1 $3d^5 4s^1/3d^6$

 If you got them all right, skip to page 48

Transition Metals

Improve your knowledge

The transition elements, Sc to Cu, **have a partially filled d sub-shell in one or more oxidation states**. Electronic configurations for Sc to Cu and Zn are given.

Key points from AS in a Week

Structure and bonding
page 18–22

Introduction to structure and bonding
pages 29–31

	Sc	Ti	V	Cr	Mn	Fe	Co	Ni	Cu	Zn
3d	1	2	3	5	5	6	7	8	10	10
4s	2	2	2	1	2	2	2	2	1	2

Note that Cr and Cu have only **one** 4s electron.
Zinc is **not** a transition metal as it has **10 d electrons in all oxidation states**.

To work out electronic structures of ions remember the **4s sub-shell empties first**.

First octet, half-filled d sub-shell (d^5) and full d sub-shell (d^{10}) show a degree of **stability**. Sc^{3+}, $3s^2 3p^6$; Mn^{2+}, d^5; and Zn^{2+}, d^{10} are stable states.

Transition metal complexes consist of a **central metal atom or ion** surrounded by **ligands**. A **ligand** is a species with a **lone pair(s)** capable of **datively** bonding into the metal's vacant d orbitals (**co-ordinate bonding**). Ligands capable of forming more than one bond are called **polydentate**. The number of bonds around the central metal is the **co-ordination number**.

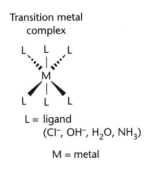

Transition metal complex

L = ligand
(Cl^-, OH^-, H_2O, NH_3)

M = metal

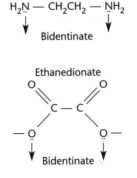

Ethandiamine

$H_2N — CH_2CH_2 — NH_2$

Bidentinate

Ethanedionate

Bidentinate

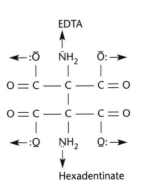

EDTA

Hexadentinate

43

Complexes are named according to ligand(s) (number and type alphabetically), metal and oxidation state.

$[Cr(H_2O)_6]^{3+}$ = hexaaquachromium (III) ion.
$[Fe(OH)_3(H_2O)_3]$ = triaquatrihydroxyiron (II).

3 When transition metals and their salts dissolve in acids or water they form **coloured, octahedral hexaaqua complex** ions, usually in either the +2 or +3 state, e.g. $[Fe(H_2O)_6]^{2+}$ and $[Cr(H_2O)_6]^{3+}$. **Aluminium** is also shown.

Ti^{2+}	V^{3+}	Cr^{3+}	Mn^{2+}	Fe^{2+}	Fe^{3+}	Co^{2+}	Ni^{2+}	Cu^{2+}	Zn^{2+}	Al^{3+}
Pink	Green	Violet	Pale pink	Pale green	Brown	Pink	Green	Blue	Colour-less	Colour-less

4 Addition of **bases**, such as OH^- ions for NH_3, to hexaaqua ions results in the formation of the **neutral, hydroxide precipitate**. Water ligands are **deprotonated** by the base to leave a hydroxide ligand.

$[Fe(H_2O)_6]^{3+} + 3OH^- \rightarrow Fe(OH)_3(H_2O)_3 + 3H_2O$
$[Co(H_2O)_6]^{2+} + 2OH^- \rightarrow Co(OH)_2(H_2O)4 + 2H_2O$

The hydroxides of Cr, Al and Zn undergo **further** deprotonation with **strong** base, i.e. OH^-, to form $[Cr(OH)_6]^{3-}$, $[Al(OH)_4(H_2O_2]^-$ and $Zn(OH)_4(H_2O)_2]^{2-}$ **anions** respectively. These hydroxides are therefore **amphoteric**.

$Al(OH)_3(H_2O)_3 + OH^- \rightarrow [Al(OH)_4(H_2O)_2]^- + H_2O$

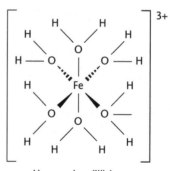

Hexaaquiron(III) ion

Deprotonation reactions are **reversed** by addition of a **strong acid**.

$Al(OH)_3(H_2O)_3 + 3H^+ \rightarrow [Al(H_2O)_6]^{3+}$

5 Hexaaqua complexes of +3 metal ions are **weak acids** due to **partial hydrolysis**. +2 hexaaqua complexes are only **very** weak acids.

$[Al(H_2O)_6]^{3+} + H_2O \rightarrow [Al(H_2O)_5(OH)]^{2+} + H_3O^+$

This difference is also shown in the reaction with **carbonate ions**: +2 ions form carbonate precipitates (MCO_3), +3 ions undergo hydrolysis to form the **hydroxide precipitate**.

$$2[Fe(H_2O)_6]^{3+} + 3CO_3^{2-} \rightarrow 2[Fe(H_2O)_3(OH)_3] + 3CO_2 + 3H_2O$$

The **difference** in acidity is due to the increased surface **charge density** on the central metal ion which strongly **polarises** electron density from the water ligands, weakening the **O–H bonds** and so facilitating the loss of H^+.

6 Addition of excess NH_3 or Cl^- can lead to **ligand exchange** reactions, i.e. one ligand is **entirely replaced** by another. The hydroxide ppt. of Cr, Cu, Co and Ni reacts with NH_3 to form octahedral, ammine complexes: $[Cr(NH_3)_6]^{3+}$, $[Cu(NH_3)_4(H_2O)_2]^{2+}$, $[Co(NH_3)_6]^{2+}$ and $[Ni(NH_3)_6]^{2+}$.

$$Ni(OH)_2(H_2O)_4 + 6NH_3 \rightarrow [Ni(NH_3)_6]^{2+} + 2OH^- + 4H_2O$$

The hexaaqua complexes of Cu and Co form the tetrahedral complexes $[CoCl_4]^{2-}$ and $[CuCl_4]^{2-}$ with excess Cl^- ligands.

$$[Co(H_2O)_6]^{2+} + 4Cl^- \rightarrow [CoCl_4]^{2-} + 6H_2O$$

Tetrahedral cobalt chloride anion

7 Transition-metal compounds are coloured due to partially filled d orbitals. d electrons absorb particular frequencies of visible light and are promoted to higher energy levels. The reflected light gives rise to the colour. Some colours of metal compounds are given below.

Complex	Cr^{3+}	Mn^{2+}	Fe^{2+}	Fe^{3+}	Co^{2+}	Ni^{2+}	Cu^{2+}	Zn^{2+}	Al^{3+}
Hydroxide ppt	Green	Beige	Pale green	Brown	Blue	Green	Blue	White	White
Hydroxide anion	Deep green							Colourless	Colourless
Ammine	Violet				Pale yellow	Blue	Deep blue		
Chloride anion				Blue		Yellow			

Colour depends on: the metal, oxidation state, ligand, and co-ordination number.

Transition Metals

Transition elements show **variable oxidation states** due to the relative ease of removing successive d electrons. Most elements show relatively stable states, +2 (Mn, Co, Ni, Cu) or +3 (Cr, Fe, Sc). Oxidation states **higher** than the stable state tend to be **oxidising** and states **lower** than the stable state tend to be **reducing** and are often themselves **oxidised by air**.

The **manganate (VII)** ion and the **dichromate (VI)** ions act as strong oxidising agents, reacting with reducing agents such as Fe^{2+} and ethanedioate.

$$MnO_4^- + 8H^+ + 5Fe^{2+} \rightarrow Mn^{2+} + 4H_2O + 5Fe^{3+}$$
$$Cr_2O_7^{2-} + 14H^+ + 3C_2O_4^{2-} \rightarrow 2Cr^{3+} + 7H_2O + 3CO_2$$

These reactions are often used in radox titrations.

Vanadium shows four oxidation states which can be chemically interconverted.

Many transition metals and compounds act as **catalysts** and are widely used in industrial processes. Catalysts **increase the rate** of reaction by offering an **alternative** chemical route of **lower activation energy**. Catalysts do **not** affect the yield.

Heterogeneous catalysts are those in a **different phase** to the reactants, usually solid catalyst and gaseous reactants. These catalysts work due to **vacant d orbitals** allowing 'temporary' bonding at their surface.

Contact process:	$SO_2 + V_2O_5 \rightarrow 2VO_2 + SO_3$	
	$2VO_2 + {}^1/_2O_2 \rightarrow V_2O_5$	(catalyst regenerated)
Haber process:	$N_2 + 3H_2 \rightarrow 2NH_3$	(Fe catalyst)
Hydrogenation:	$R_2C=CR_2 + H_2 \rightarrow R_2CH–CHR_2$	(Ni catalyst)

Homogeneous catalysts are those in the same phase as the reactants, usually in solution. These catalysts work due to **variable oxidation state**.

$S_2O_8^{2-} + 2I^- \rightarrow 2SO_4^{2-} + I_2$ Uncatalysed reaction: 2 **anions** collide.

$S_2O_8^{2-} + 2Fe^{2+} \rightarrow 2SO_4^{2-} + 2Fe^{3+}$ **then** $2I^- + 2Fe^{3+} \rightarrow I_2 + 2Fe^{2+}$

Catalysed reaction: anion and cation collide which is **easier**. Fe changes oxidation state.

Make sure that you understand the following key terms:

partially filled d orbitals / complex / ligand / coloured compounds / hexaaqua ions / hydroxide ppt. / deprotonation / ligand exchange / difference between +3 and +2 metal ions / variable oxidation state / heterogeneous and homogeneous catalysis

Transition Metals

Use your knowledge

1 a) Define and give three characteristic properties of transition elements.

Learn this!

b) Give the outer shell configuration of Cr^{3+}, V and Ni^{2+}.

4s electrons lost first

c) Why is +2 the most stable oxidation state of Mn but not of Fe?

Mn^{2+} is $3d^5$

d) Predict and explain the most stable oxidation state of scandium.

Full octet

2 Chronium sulphate dissolves in water to form a violet solutions, A. When conc. NH_3 is added dropwise to a portion of A, a green ppt. B forms, which dissolves in excess NH_3 forming a violet solution, C. Addition of aqueous NaOH to A also forms a green ppt., D, which dissolves in excess NaOH to form a deep green solution, E. Give the formula A to E and name A and C.

NaOH leads to hydrolysis only ... NH_3 can also lead to ligand exchange

3 a) $Al(OH)_3$ is amphoteric. Explain, with equations, what this means.

Acidic and basic

b) When green iron(II) hydroxide precipitate was allowed to stand in air it was observed to turn brown. Explain this observation and why it occurs.

Air can oxidise to a more stable state

c) Explain why a solution of aqueous Al^{3+} ions is acidic.

Aluminium is $3+$ and hydrolyses

4 a) Fe^{2+} ions catalyse the reaction $S_2O_8^{2-} + 2I- \rightarrow 2SO_4^{2-} + I_2$. What type of catalysis is this and on what feature of the transition metal does it rely?

$Fe^{2+}/I^-/S_2O_8^{2-}$ all aqueous ... Fe can easily gain and lose electrons

b) Give equations for and explain why the catalysed reactions are faster.

10 minutes

Test your knowledge

1 What structural feature of haloalkanes leads to their reactivity?

2 What type of mechanism do primary haloalkanes undergo with aqueous hydroxide ions?

3 What type of mechanism do tertiary haloalkanes undergo with aqueous hydroxide ions (as distinct to primary haloalkanes)?

4 Give the reagents and conditions required to convert bromoethane into a Grignard reagent.

5 Name the compound formed when the Grignard reagent formed in Question 4 reacts with methanal followed by dilute H_2SO_4.

6 What type of compound is formed by the reaction of Grignard reagents with carbon dioxide?

Answers

6 Carboxylic acids
5 Propan-1-ol
4 Mg metal, dry ether, iodine catalyst
3 First-order nucleophilic substitution (SN_1)
2 Second-order nucleophilic substitution (SN_2)
1 Polar carbon–halogen bond

 If you got them all right, skip to page 53

Haloalkanes 2

10 minutes

Improve your knowledge

1. Haloalkanes are saturated hydrocarbons containing a polar, covalent carbon-halogen bond. The $\delta+$ carbon atom is susceptible to nucleophilic substitution to form the compounds shown:

    OH^- ions (from warm NaOH(aq)) produce alcohols.
    CN^- ions (from hot KCH in ethanol) produce nitriles.
    NH_3 molecules (hot, alcoholic NH_3) produce amines.

2. Primary haloalkanes undergo second-order nucleophilic substitution, SN_2. The nucleophile attacks the positive carbon opposite the halogen and the carbon-halogen bond breaks simultaneously, via a 5-coordinate transition state. Because both species are present in a single step, the reaction rate is first order with respect to each and so second order overall (rate = $k[RX][Nu]$).

    HO ---- C ---- X

    Transition state

3. Tertiary haloalkanes undergo first-order substitution, SN_1. Bulky groups prevent attack by the nucleophile of the carbon atom, therefore SN_2 is not possible. However, positive inductive effects from the alkyl groups stabilise cations formed by the breaking of the carbon–halogen bond. The nucleophile then bonds with the carbocation in a second step. The first step involves endothermic bond breaking and so is slow, whereas the second involves bond formation and is therefore fast. Thus the first step controls the overall rate of reaction and because only the haloalkane is present in this step the rate of reaction is first order with respect to the haloalkane only (rate = $k[RX]$).

Key points from
AS in a Week

Basic organic
chemistry –
structure, naming
and isomerism of
organic molecules,
homologous series
pages 60–62

Haloalkanes
pages 72–76

Kinetics
pages 37–41

Haloalkanes 2

Carbocation
Intermediate

Because the nucleophile can attack the **planar** cation **equally** from above or below, the product formed is often **racemic**.

4 Haloalkanes react with magnesium metal in a dry ether solvent and an iodine catalyst to produce **Grignard** reagents.

$$CH_3Br + Mg \rightarrow CH_3MgBr \quad \text{methylmagnesium bromide}$$

Grignard reagents contain a highly polar C–Mg bond as shown and act as a 'source' of **alkyl anions** in nucleophilic reactions in which they **extend the carbon chain**.

5 Grignard reagents react with **carbonyls** in dry ether to produce **alcohols**. The initial product formed is a **salt**, which is converted to the corresponding alcohol by dilute H_2SO_4 or HCl. The carbonyl determines the structure of the alcohol:

Grignard + **Methanal** → **Primary** alcohol
Grignard + **Other Aldehyde** → **Secondary** alcohol
Grignard + **Ketone** → **Tertiary** alcohol

$$R-MgX + \overset{H}{\underset{H}{\diagdown}}C=O \longrightarrow R-\overset{\overset{H}{|}}{\underset{\underset{H}{|}}{C}}-OH + Mg \times OH$$

Methanol → Primary alcohol

$$R-MgX + \overset{R'}{\underset{H}{\diagdown}}C=O \longrightarrow R-\overset{\overset{R'}{|}}{\underset{\underset{H}{|}}{C}}-OH + Mg \times OH$$

Aldehyde → Secondary alcohol

$$R-MgX + \overset{R'}{\underset{R^2}{\diagdown}}C=O \longrightarrow R-\overset{\overset{R'}{|}}{\underset{\underset{R^2}{|}}{C}}-OH + Mg \times OH$$

Ketone → Tertiary alcohol

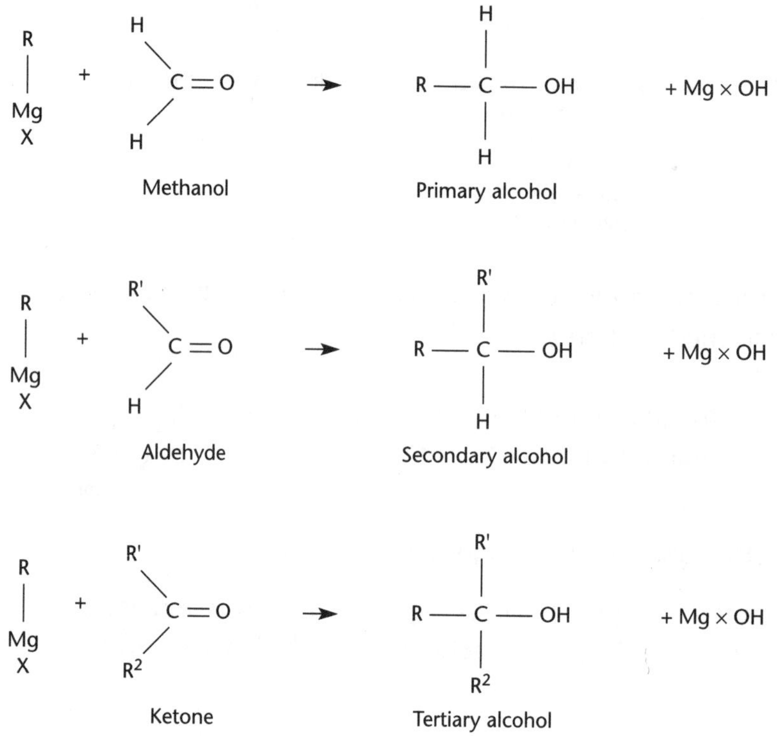

6 Grignard reagents react with CO_2 to form **carboxylic acids** and with **water** to form **alcohols**, again following acidification.

Make sure that you:

understand SN_1 and SN_2 mechanisms / can recall the reactions and conditions of haloalkanes with NaOH, KCN, NH_3 / recall the formation and importance of Grignard reagents and the products they form with carbonyls, CO_2 and water

Haloalkanes 2

Use your knowledge

1 A and B are isomeric haloalkanes with molecular formula $C_7H_{15}Br$. Both undergo nucleophilic substitution with warm aqueous NaOH to produce alcohols C and D respectively. A is un-branched and its substitution is found to be first order with respect to the haloalkane and the hydroxide ions. B is branched and its substitution reaction is found to be first order with respect to the haloalkane and zero order with respect to the hydroxide ions.

a) Give the structure and name of A.

b) Draw the mechanism for the reaction between A and OH⁻ ions. Explain why it is first order with respect to both species.

c) Suggest a structure for B and give the mechanism for its substitution by hydroxide ions.

d) Explain why the mechanisms for A and B are different.

e) Explain why although B contains a chiral carbon atom and displays optical activity, D is found to be optically inactive.

2 The molecule 2-methylhexan-2-one can be synthesised using a Grignard reagent under appropriate conditions.

a) Give the reagents and conditions necessary to prepare propyl magnesium bromide.

b) Give the structure of hexan-2-one and outline how it could be prepared from propylmagnesium bromide.

c) What feature of Grignard reagents leads to their reactivity?

d) Grignard reagents are used to extend to the carbon chain. Despite the expense, explain why they are often preferred to reaction with cyanide ions.

Must be primary

Single step

OH⁻ ions cannot directly attack haloalkane

Stability of cations

Cation intermediate is planar

Mg metal

Grignard + aldehyde

Highly polar C-Mg bond

Fewer steps

15 minutes

Test your knowledge

1 Give the formula of benzene and describe the unique structural feature it contains.

2 How is the enthalpy change of hydrogentation unusual for benzene and what does it suggest?

3 Write an equation to show the formation of the NO_2^+ species in the nitration of benzene and name the mechanism.

4 Suggest the reagents required to introduce a methyl group onto the benzene ring in Friedel–Crafts alkylation.

5 What other group can be introduced by Friedel–Crafts reactions?

6 Identify the product when methylbenzene is completely oxidised using alkaline potassium manganate.

7 Name the reaction between benzene and SO_3.

8 Give the structural formula of phenol.

Answers

8 C_6H_5OH
7 Sulphonation
6 Benzoic acid/C_6H_5COOH
5 Acyl
4 $CH_3Cl + AlCl_3$
3 $2H_2SO_4 + HNO_3 \rightarrow 2HSO_4^- + NO_2^+ + H_3O^+$/electrophilic substitution
2 It is less than expected/thermodynamic stability.
1 C_6H_6/three delocalised pi-bonds

✔ **If you got them all right, skip to page 58**

Aromatic Compounds

Improve your knowledge

1 Benzene, C_6H_6, is a flat, six-carbon ring containing three pi bonds. All six C–C bonds are of equal length, indicating that the pi-electrons are delocalised, above and below the ring, which leads to thermodynamic stability.

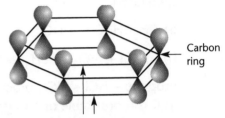

Carbon ring

Pi electrons (bonds) delocalised above and below ring

2 The stability of benzene is shown by hydrogenation reactions. The energy released when benzene is hydrogenated is not three times that when cyclohexene is hydrogenated, as would be predicted if fixed C=C bonds were present.

Cyclohexene: $C_6H_{10} + H_2 \rightarrow C_6H_{12}$ $\Delta H = -120$ kJ mol^{-1}
Benzene: $C_6H_6 + 3H_2 \rightarrow C_6H_{12}$ $\Delta H = -208$ kJ mol^{-1}
not 360 kJ mol^{-1}.

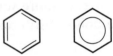

Benzene can be represented in two ways

3 Benzene undergoes electrophilic substitution. Addition doesn't occur across pi bonds as this would permanently disrupt delocalisation, reducing stability. For example, nitration, by reaction with conc. H_2SO_4 and conc. HNO_3 at 50–60°C (poly-substitution may occur at higher temperatures). Aromatic nitro-compounds are widely used in the production of explosives and dyes.

Formation of electrophile: $2H_2SO_4 + HNO_3 \rightarrow NO_2^+ + 2HSO_4^- + H_3O^+$
Overall: $C_6H_6 + HNO_3 \rightarrow C_6H_5NO_2 + H_2O$

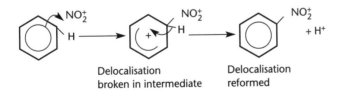

Delocalisation broken in intermediate

Delocalisation reformed

Nitrobenzene may be reduced to phenylamine by tin metal and conc. HCl.

$C_6H_5NO_2 + 6[H] \rightarrow C_6H_5NH_2 + 2H_2O$

Aromatic Compounds

4 Friedel–Crafts alkylation substitutes alkyl groups, R, onto the ring by electrophilic substitution. The electrophile, R^+, can be generated as shown.

a) $CH_3CH=CH_2 + HCl + AlCl_3 \rightarrow CH_3C^+HCH_3 + AlCl_4^-$

b) $CH_3Cl + AlCl_3 \rightarrow CH_3^+ + AlCl_4^-$ ($AlCl_3$ is a catalyst)

Propene + HCl + $AlCl_3$ produces the 2° rather than the 1° cation as it is more stable due to greater positive inductive effects. Thus the product is methylethylbenzene (cumene), not propylbenzene.

Alkyl groups on the benzene ring increase negative charge in the delocalised ring due to positive inductive effects. This makes alkyl benzenes more susceptible to electrophilic substitution, leading to poly-substitution.

Methylbenzene

Methyelbenzene

5 Friedel–Crafts acylation substitutes acyl, RCO, groups by electrophilic substitution. The electrophile, RC^+O, is formed by reacting acyl chlorides and $AlCl_3$.

a) $CH_3COCl + AlCl_3 \rightarrow CH_3C^+O + AlCl_4^-$

b) $C_6H_5COCl + AlCl_3 \rightarrow C_6H_5C^+O + AlCl_4^-$

Phenylethanone

6 Benzene compounds containing side chains are converted to benzoic acid when strongly oxidised, such as with alkaline potassium manganate. If the side chain contains more than one carbon then they are released as CO_2.

Benzphenone

Benzoic acid

Aromatic Compounds

7 Benzene undergoes sulphotation with SO_3 in conc. H_2SO_3.

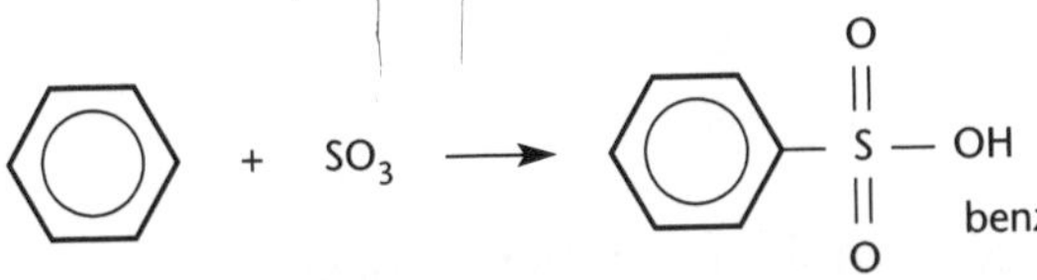

benzsulphonic acid

8 Phenol, C_6H_5OH, is weakly acidic due to the withdrawal of electron density from the O–H bond. This accounts for the reaction with bases to produce phenoxide salts. The increased charge density in the delocalised ring increases susceptibility to electrophilic substitution and phenols react with bromine at room temperature without a catalyst, unlike benzene. Phenol also undergoes acylation with acid chlorides to produce aromatic esters. These esters are not readily formed using the corresponding carboxylic acid.

$$C_6H_5OH + NaOH \rightarrow C_6H_5O^+Na + H_2O$$

$$C_6H_5OH + 3Br_2 \rightarrow C_6H_2Br_3OH + 3HBr$$

$$C_6H_5OH + CH_3COCl \rightarrow C_6H_5OOCCH_3 + 3HBr$$

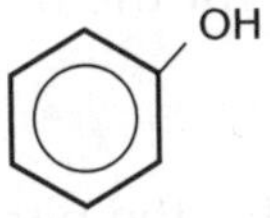

Phenol

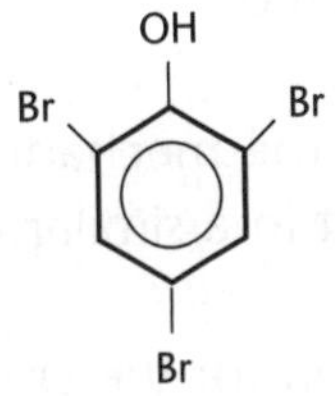

2, 4, 6 – tribromophenol

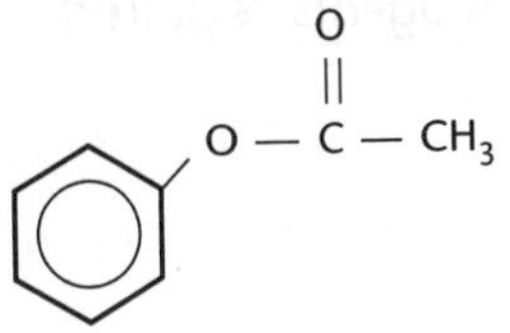

Phenyl ethanoate

Make sure you are familiar with the following terms:

benzene / delocalisation stability / nitration / electrophilic substitution / sulphonation / Friedel–Crafts alkylation and acylation / oxidation of aromatic side chains / reactions of phenol

Aromatic Compounds

Use your knowledge

1 a) Describe the structure and bonding in benzene.

Ring structure / pi bonds

b) When cyclohexene is hydrogenated $\Delta H = -120$ kJ mol^{-1} but when benzene is hydrogenated $\Delta H = -204$ kJ mol^{-1}, despite the presence of three pi bonds. Write an equation for the hydrogenation of benzene and explain the ΔH value.

Addition reaction saturates pi bonds

2 a) Give an equation for the nitration of benzene and the reagents and conditions required.

Learn this

b) Name and give the mechanism for nitration and explain why benzene does not usually react in a similar manner to alkenes.

Addition reactions reduces stability

c) Write an equation for the conversion of nitrobenzene to phenylamine and give the reagents required.

Phenylamine is $C_6H_5NH_2$

Acids, Esters and Acid Chlorides

Test your knowledge

1 Write an equation for the reaction between ethanoic acid and sodium carbonate.

2 Suggest a reagent to convert propanoic acid to propanol.

3 Give the type and name of the product when ethanoic acid is heated under reflux with methanol with conc. sulphuric acid.

4 What characteristic property do esters have and what commercial use does this lead to?

5 What two products are formed when ethyl butanoate is heated strongly in aqueous NaOH?

6 What is the commercial application of the reaction described in Question 5?

7 What type of compound is formed from di-carboxylic acids and diols, and what are they used for?

8 Suggest a reagent for preparing ethanoyl chloride from ethanoic acid.

9 Name the products when ethanoyl chloride reacts with water.

10 Name the mechanism for the reaction in Question 9.

11 Give the structural formula of ethanoic anhydride.

Answers

1 $CH_3CO_2H + NaOH \rightarrow CH_3COONa + H_2O$ **2** $LiAlH_4$ **3** Ester/methyl ethanoate **4** Strong odour/food flavourings or perfumes. **5** Ethanol and sodium butanoate **6** Making soap. **7** Polyesters/synthetic fabrics **8** PCl_5 or $SOCl_2$ **9** Ethanoic acid and hydrogen chloride **10** Nucleophilic addition–elimination **11** $(CH_3CO)_2O$

✔ If you got them all right, skip to page 63

Acids, Esters and Acid Chlorides

30 minutes

Improve your knowledge

1 Carboxylic acids, RCOOH, are **weak** acids and react accordingly.

Dissociation in water: $C_6H_5COOH + H_2O \rightarrow CH_3COO^- + H_3O^+$
$\qquad\qquad\qquad\qquad\qquad$ **ethanoate ion**

Neutralisation: $HCOOH + NaOH \rightarrow HCOO^-Na^+$ (reversed by strong acid)
$\qquad\qquad\qquad\qquad\quad$ **sodium methanoate (salt)**

With **carbonate:** $C_2H_5COOH + NaHCO_3 \rightarrow C_2H_5COO^-Na^+ + H_2O + CO_2$
(CO_2 forms a milky **precipitate** with lime water. This is a chemical **test** for acids.)

> Key points from
> AS in a Week
> ___
> Alcohols, carbonyls
> and acids   page 77

2 Carboxylic acids are **reduced** to **aldehydes** and then **1° alcohols** by $LiAlH_4$.

$$C_2H_5COOH + 2[H] \rightarrow H_2O + C_2H_5CHO + 2[H] \rightarrow C_2H_5CH_2OH$$

3 Carboxylic acids react with **alcohols** when **heated under reflux** with a **conc. H_2SO_4** catalyst to form **esters**. This is a **reversible** reaction. Esters are named using the alkyl prefix from the alcohol and the acid ion suffix.

$$\begin{array}{c} O \\ \parallel \\ CH_3 - C \\ \diagdown \\ O - CH_3 \end{array}$$

$\underbrace{}_{\text{Ethanoate}}$ $\underbrace{}_{\text{Methyl}}$

$$CH_3COOH + CH_3OH \rightarrow H_2O + CH_3COOCH_3 \text{ methyl ethanoate}$$

4 Esters are formed and named from their parent acids and alcohols as shown above. They have strong odours and aliphatic esters are **sweet smelling** and used in **food flavourings** and **perfumes**. Aromatic esters often smell 'antiseptic'.

5 Esters are **hydrolysed** by dilute acid or alkali when heated under reflux.

Acid: $CH_3COOCH_3 + H_2O \rightarrow CH_3COOH + CH_3OH$ (reversible, low yield)
Alkali: $C_2H_5COOCH_3 + NaOH \rightarrow C_2H_5COONa + CH_3OH$ (non-reversible)

6 The alkaline hydrolysis of natural triglyceride esters of stearic acid leads to sodium stearate, which is used as hand soap. The reaction is known as saponification.

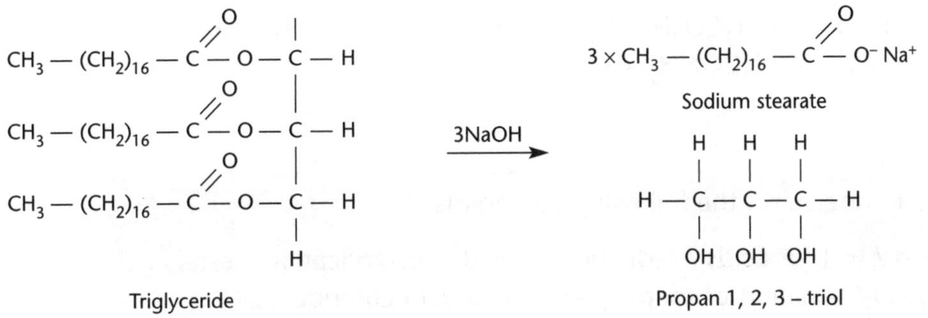

Triglyceride

$3NaOH$

$3 \times CH_3 - (CH_2)_{16} - C \overset{O}{=} O^- Na^+$

Sodium stearate

Propan 1, 2, 3 – triol

7 Di-acids and diols esterify to form polyesters used as synthetic fibres.

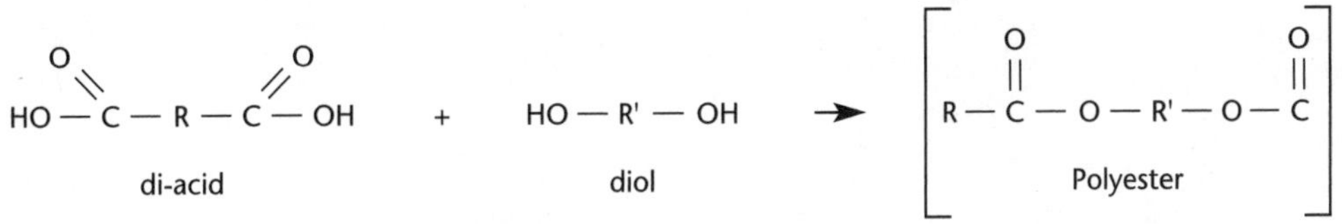

di-acid   diol   Polyester

8 Acid chlorides, RCOCl, are formed by chlorination of carboxylic acids.

$CH_3COOH + PCl_5 \rightarrow CH_3COCl$ (**ethanoyl chloride**) $+ HCl + POCl_3$
$C_6H_5COOH + SOCl_2 \rightarrow C_6H_5COCl$ (**benzoyl chloride**) $+ HCl + SO_2$

9 Acid chlorides are extremely reactive due to a highly δ+ carbon atom bonded to electronegative O and Cl. They readily react at room temperature with nucleophiles, H_2O, ROH, NH_3 and RNH_2, which are **acylated**.

$CH_3COCl + HOH \rightarrow CH_3COOH + HCl$ hydrolysis
$C_6H_5COCl + HOCH_3 \rightarrow C_6H_5COOCH_3 + HCl$ esterification
$CH_3COCl + HNH_2 \rightarrow CH_3CONH_2 + HCl$ forming of amide
$CH_3COCl + HNRH \rightarrow CH_3CONRH + HCl$ forming substituted amides

10 The reactions above occur by **nucleophilic addition elimination**.

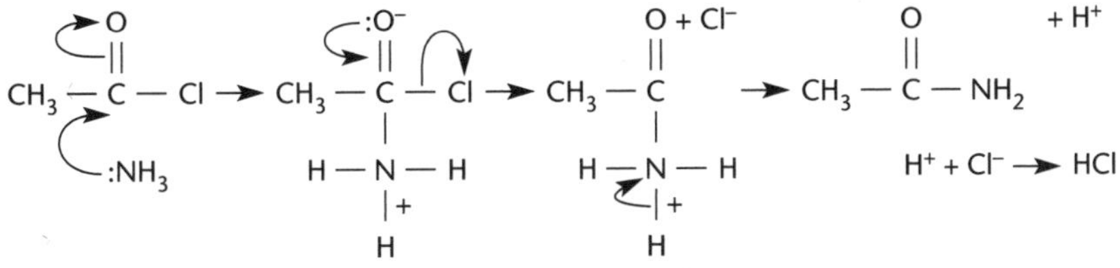

Acids, Esters and Acid Chlorides

 Acid anhydrides react according to the same mechanism and in a similar manner to acid chlorides but instead of HCl carboxylic acids are produced.

$$(CH_3CO)_2O + H_2O \rightarrow 2CH_3CO_2H \quad \text{hydrolysis of ethanoic anhydride}$$
$$(CH_3CO)_2O + NH_3 \rightarrow CH_3CONH_2 + CH_3CO_2H$$

Make sure you understand the following key points:

carboxylic acid / test for acids / reduction of acids / esterification / esters / ester hydrolysis / formation of soap / polyesters / acid chlorides / nucleophilic addition–elimination mechanism and reactions

Acids, Esters and Acid Chlorides

45 minutes

Use your knowledge

1 A carboxylic acid **M**, $C_3H_6O_2$, reacts as follows: + $LiALH_4$ → **N**; + NaOH → **O**; + CH_3OH/conc. H_2SO_4 → **P**; + PCl_5 → **Q** (organic) and **R** (inorganic) and **S** (gas); + $NaHCO_3$ → **O** and **T** (gas).

a) Identify N, O, P and the type of reactions occurring in their formation.

b) Identify gas T and suggest how its identity could be confirmed.

c) Give an equation for the formation of Q, R and S. Identify Q, R and S.

d) Name and give the structure of M.

2 The formula $C_4H_8O_2$ represents four isomeric esters A, B, C and D. When A and B are heated in dilute H_2SO_4 they form propanoic acid and E, and propanol and F respectively. C forms ethanol and G when heated with dilute NaOH.

a) Give the name and structure of A, B and C and identify E, F and G.

b) Give the structure of D and name the carboxylic acid and alcohol from which it could be made.

3 a) Ethanoyl chloride reacts vigorously with water. Give the equation and mechanism for this reaction and the type of reaction.

b) Explain why acid chlorides are so reactive.

c) Suggest two reactants necessary to produce the molecule $C_6H_5CON(CH_3)H$. What type of compound is this?

d) Give an equation for the reaction between ethanoic anhydride and methanol.

LiALH$_4$ is a reducing agent

NaOH is a base

P is sweet smelling

PCl$_5$ chlorinates OH groups

Acid + carbonate

Learn this

Esters have the general formula R$_1$COOR$_2$

Acid hydrolysis forms acids R$_1$COOH and alcohols R$_2$OH

Alkaline hydrolysis leads to salts and alcohols

D has a branched chain

Reaction with water forms an acid

Nucleophiles attack δ+ carbons

HCl is eliminated to form C$_6$H$_5$CON(CH$_3$)H

Similar reaction to acid chlorides, ethanoic acid formed instead of HCl

10 minutes

Test your knowledge

1 Name the two types of carbonyls.

2 Name the product when propanone is reduced.

3 Give the name of the product when ethanal is oxidised.

4 By what mechanism does ethanal react with HCN?

5 Explain what is meant by a chiral carbon atom.

6 Explain what is meant by a racemic mixture.

7 Describe what is observed when Fehling's solution is added to propanal.

8 Give the reagent and observation for a chemical test for all carbonyls.

Answers

1 Aldehydes and ketones
2 Propan-2-ol
3 Ethanoic acid
4 Nucleophilic addition
5 A carbon atom surrounded by four different groups.
6 A 50:50 molar mixture of two opposite enantiomers.
7 Blue solution forms brick-red ppt.
8 2,4-DNP gives orange/red ppt.

 If you got them all right, skip to page 67

Carbonyls and Optical Isomerism

10 minutes

Improve your knowledge

1 Carbonyls (C=O) exist as aldehydes, RCHO, or ketones, R_1COR_2.

$$\underset{CH_3}{\overset{H}{\diagdown}}C = O \quad \text{"Ethanal"}$$

Key points from AS in a Week
Alcohols, carbonyls and acids   page 77

2 Aldehydes and ketones can be prepared by **oxidation** of **1°** and **2° alcohols** respectively. They are reduced to alcohols by $LiAlH_4$ or $NaBH_4$.

$$\underset{C_2H_5}{\overset{CH_3}{\diagdown}}C = O \quad \text{Butanone}$$

$CH_3CH_2OH + [O] \rightarrow CH_3CHO + H_2O$ ethanol to ethanal
$CH_3COCH_3 + 2[H] \rightarrow CH_3CH_2(OH)CH_3$ propanone to propan-2-ol

3 Aldehydes are **oxidised** to carboxylic acids by $KMnO_4$ or $K_2Cr_2O_7$ in dil. H_2SO_4 (**purple** to **colourless** and **orange** to **green**). Ketones do **not** oxidise.

$$R - \overset{\overset{O}{\|}}{\underset{\underset{H}{|}}{C}} \rightarrow R - \overset{\overset{O}{\|}}{C} - OH$$

$CH_3CHO + [O] \rightarrow CH_3COOH$ ethanal to ethanoic acid

4 Carbonyls undergo **nucleophilic addition** with **HCN** in the presence of OH^- ions, giving **2-hydroxynitriles**. The nucleophile is the CN^- ion, generated by the dissociation of the HCN. This dissociation is favoured by alkaline conditions.

$$H - CN \rightleftharpoons H^+ + :CN^-$$

$$\underset{H}{\overset{CH_3}{\diagdown}}C = O \longrightarrow CH_3 - \overset{\overset{H^+ \curvearrowright :O^-}{\|}}{\underset{\underset{CN}{|}}{C}} - H \longrightarrow CH_3 - \overset{\overset{OH}{|}}{\underset{\underset{H}{|}}{C}} - H$$

:CN⁻

Nucleophile can attack from above and below

2-Hydoxy propanitrile

5 Hydroxynitriles are often **chiral**. A chiral molecule is one where a carbon atom is surrounded by **four different groups**. This leads to the formation of two **non-superimposable mirror images** called **enantiomers**. Each enantiomer rotates the **plane** of **plane-polarised** light in **opposite** directions. Although the functional

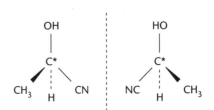

Carbonyls and Optical Isomerism

group chemistry of the two enatiomers is identical, they can have very different **biological activity**. Therefore it is very important that pharmaceuticals are synthesised so that the correct enantiomer is produced.

6 The product of the reaction between carbonyls and HCN is often optically inactive, despite containing chiral molecules. This is because carbonyls are **flat** so the CN^- ion can attack equally from **above and below**, forming a 50:50 mixture of both isomers. This is called a **racemic mixture** and because the activity of the two isomers cancels each other out the mixture is **optically inactive**.

7 Aldehydes can be identified by **Tollen's reagent** or **Fehling's solution**, which are mild oxidants, converting the aldehyde to the carboxylic acid or acid ion if in alkaline solution.

Tollen's reagent ($AgNO_3/NH_3$) colourless $Ag^+ \rightarrow$ silver 'mirror' of Ag^0

Fehling's solution ($Cu^{2+}(aq)$) blue $Cu^{2+} \rightarrow$ brick-red ppt. of Cu_2O

8 Carbonyls are detected using **2,4-dinitrophenylhydrazine (2,4-DNP)** solution to produce a **red/orange ppt**. The carbonyl is identified by recrystallisating the ppt. and comparing its melting point with literature values. The ppt. is first re-crystallised from a minimum of **ethanol**, filtered **under suction** and **dried**.

2, 4 DNP

Make sure you are familiar with the following key points:

carbonyl / aldehyde / ketone / carbonyl oxidation and reduction / nucleophilic addition with HCN / optical isomerism / racemic mixture / Fehling's and Tollen's / 2,4-DNP / recrystallisation and melting-point determination

30 minutes

Use your knowledge

1 a) Give an equation for the reaction between ethanal and HCN and name the product. Indicate the chiral carbon with an asterisk.

b) Name and draw the mechanism for this reaction.

c) Explain why the product is optically inactive.

2 An unknown organic molecule A, $C_4H_{10}O$, reacted with an aqueous solution of $K_2Cr_2O_7$ and dilute H_2SO_4 to form B initially and then C, on further reaction. B reacted with a solution of $AgNO_3$ in aqueous NH_3 to form a silver deposit, D, and an organic product E. B also reacts with HCN to produce F.

a) Identify A–F.

b) What type of reaction is occurring in the formation of B and C and what would be observed?

c) G, an isomer of B, also reacts with HCN, does not react with $AgNO_3$ and NH_3, and reacts with $NaBH_4$ to form H. Identify G and H and suggest and explain what would be observed when G is reacted with acidified $K_2Cr_2O_7$.

d) Describe how both B and G could tested for and briefly how this test could be used to identify precisely either of the two compounds.

HCN adds to ethanal

Chiral carbons have four different groups attached

CN^- is the nucleophile

It attacks the positive C-atom

Both isomers formed equally

A is a 1° alcohol

B is a carbonyl

Ag/NH_3 is Tollen's reagent

F is a 2-hydroxynitrile

Learn this

B is a ketone

2,4-DNP ppt.

Organic Nitrogen Compounds

15 minutes

Test your knowledge

1. Give the formula of ethylamine and explain why it is a base.

2. Give the product formula and name when ethylamine reacts with HCl.

3. By what mechanism do amines react with haloalkanes?

4. Give the general formula of amino acids.

5. What is the product when phenylamine reacts with $NaNO_2$ and conc. HCl at cold temperatures?

6. Give an equation for the reduction of propanitrile.

7. What type of organic compound is formed when nitriles undergo hydrolysis?

8. Name the molecule $C_2H_5CONH_2$.

9. Give the name and formula of a molecule that could be dehydrated to form the molecule in Question 8.

10. Give the formula and name of the product when CH_3CONH_2 reacts with $LiAlH_4$.

11. Give the product formula and reaction name when $C_2H_5CONH_2$ reacts with Br_2 and conc. NaOH.

12. Name a commercial polyamide.

Answers

1 $C_2H_5NH_2$/lone pair on N for bonding 2 $C_2H_5NH_3Cl$/ethylammonium chloride 3 Nucleophilic substitution 4 $H_2N-R-COOH$ 5 Benzenediazonium chloride 6 $C_2H_5CN + 4[H] \rightarrow C_2H_5CH_2NH_2$ 7 Carboxylic acids 8 Propanamide 9 $C_2H_5COO-NH_4$ 10 $C_2H_5NH_2$/propylamine 11 $C_2H_5NH_2$/Hofmann degradation 12 Nylon 66

✔ **If you got them all right, skip to page 73**

68

30 minutes

Improve your knowledge

1 Amines, RNH_2, are **bases** as the **lone pair** on the N atom accepts H^+ ions. Amines are stronger bases than ammonia, due to **positive inductive** effects from **alkyl groups** making the lone pair **more available** for bonding. **Aromatic** amines are **weaker** bases than ammonia due to the **electron-withdrawing** effects of the ring which **reduces** lone pair availability.

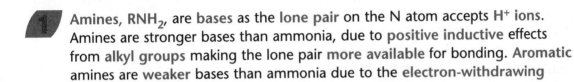

R — N̈ — H	H — N̈ — H	⬡—N̈ — H
H	H	H
Amine	Ammonia	Phenylamine

2 Amines often have poor solubility in water but react with acids to form **ionic salts**, which are often soluble.

$$C_2H_5NH_2 + HCl \rightarrow C_2H_5NH_3^+Cl^- \quad \text{ethylammonium chloride}$$

3 Amines are **nucleophiles** due to their **lone pair** and react with haloalkanes by repeated nucleophilic substitution reactions to form to **secondary** and **tertiary** amines, R_2NH and R_3N, and **quaternary ammonium salts**, $R_4N^+X^-$. Long-chain salts are used as **cationic detergents**.

$$CH_3Cl + C_2H_5NH_2 \rightarrow HCl + C_2H_5N(H)CH_3 \quad \text{ethylmethylamine}$$

$$2C_2H_5Br + (C_2H_5)_2NH \rightarrow HBr + (C_2H_5)_4N^+Br^- \quad \text{tetraethylammonium bromide}$$

Organic Nitrogen Compounds

4 Amino acids contain amine **and** carboxylic acid groups, e.g. aminoethanoic (glycine) acid, H_2N-CH_2-COOH, and react to form salts with both acids and bases (i.e. $H_2N-CH_2-COO^-\ ^+M$ with bases and $X^-H_3^+N-CH_2-COOH$ with acids). In neutral solutions and the pure state amino acids exist as **Zwitterions**, $H_3^+N-CH_2-COO^-$. The associated ionic bonding in the pure state gives rise to relatively high melting-point solid structures.

Amino acid

Zwitterion

5 Phenylamine reacts with sodium nitrite, $NaNO_2$, and conc. HCl (producing nitrous acid, HONO, *in situ*) at **temperatures < 10°C**, to form **benzene diazonium chloride**, $C_6H_5-N^+\equiv N\ Cl^-$.

$$C_6H_5NH_2 + NaNO_2 + 2HCl \rightarrow C_6H_5N^+NCl^- + NaCl + 2H_2O$$

The reactive diazonium salt undergoes coupling reactions with phenols, naphthols and aromatic amines to form **diazo compounds**. They have distinctive red, orange or yellow colours and are used as **commercial dyes**.

At temperatures above 10°C the diazonium intermediate undergoes hydrolysis, releasing nitrogen gas.

6 Nitriles, **RCN**, can be **reduced** to form **amines** using $LiAlH_4$ in dry ether (the carbon-nitrogen triple bond is saturated). This is a better method for preparing amines than reaction between ammonia and haloalkanes because **no further** reactions occur, thus leading to a better yield.

$$C_2H_5CN + 4[H] \rightarrow C_2H_5CH_2NH_2 \quad \text{propanenitrile to propylamine}$$

7 Nitriles can be hydrolysed by boiling with dilute aqueous acids to form carboxylic acids (alkaline hydrolysis leads to the carboxylic salt).

$$C_3H_7CN + 2H_2O \rightarrow C_2H_5COOH + NH_3 \text{ propanenitrile to propanoic acid}$$

8 Amides, $RCONH_2$, are much weaker bases than amines because the lone pair on the N atom is delocalised onto the O atom and unavailable for bonding.

9 Amides can be prepared from the dehydration of ammonium salts.

$$CH_3CO_2^-NH_4^+ \rightarrow CH_3CONH_2 + H_2O \quad \text{ammonium ethanoate to ethanamide}$$

10 $LiAlH_4$ reduces amides to amines and P_2O_5 dehydrates them to nitriles.

$$CH_3CONH_2 + 4[H] \rightarrow CH_3CH_2NH_2 \text{ (amine with same no. of C atoms)}$$
$$CH_3CONH_2 \rightarrow CH_3CN + H_2O \text{ (dehydration)}$$

11 Reaction with Br_2 and conc. NaOH converts amides, $RCONH_2$, to amines, RNH_2. This reaction is the Hofmann degradation and is extremely important because the carbon skeleton is reduced by one.

$$CH_3CONH_2 + Br_2 + 4NaOH \rightarrow CH_3NH_2 + 2NaBr + Na_2CO_3 + H_2O$$
Ethanamide   Methylamine

12 1,6-hexyldiamine and 1,6-hexanedioic acid undergo a condensation reaction to form the polyamide Nylon 66, used for synthetic fibres and textiles.

Organic Nitrogen Compounds

The amide link is susceptible to hydrolysis by acids and alkalis, making polyamides less inert than polymers containing only C–C bonds in the backbone, such as polyethene. Polypeptides (proteins) are natural polyamides. Their hydrolysis by acids reforms the constituent amino acids from which they are made.

Make sure you understand the following key points:

amines / amine basicity / ammonium salts / amino acids / benzdiazonium chloride / coupling reactions / diazo compounds / nitriles / nitrile hydrolysis and reduction / amides / dehydration of ammonium salts / amide reduction and dehydration / Hofmann degradation / polyamides / nylon

45 minutes

Use your knowledge

1 a) Explain, giving an equation, why ethylamine is a base.

b) Why is ethylamine a stronger base than NH_3 but phenylamine weaker?

c) Give an equation to show the preparation of ethylamine from ammonia and bromoethane. Explain, with a mechanism, why there is a low yield.

d) Give an alternative method, reagent and equation for the preparation of ethylamine that gives a better yield. Explain why it gives a better yield.

2 Ammonium ethanoate dehydrates giving amide A, which reacts with P_2O_5 to form B, C_2H_3N, which reacts with $LiAlH_4$ to form C, C_2H_7N. Boiling in dil. H_2SO_4 converts B into D, $C_2H_4O_2$. Heating A with Br_2 in NaOH forms a suspension, E, which forms a clear solution, F, on addition of HCl.

a) Identify A–F and the type of reactions for the formation of B–D and F.

b) What is the name and synthetic importance of the reaction to form E?

c) Why is E formed as an oily suspension that dissolves in acid?

3 a) Explain why amides are weaker bases than amines.

b) The polymer Kevlar can be produced from 3-aminobenzoic acid, $H_2N-C_6H_4-COOH$. Draw the structure of the Kevlar repeat unit. What type of reaction is involved and what type of polymer is Kevlar?

c) Give the structure of the species formed by 2-aminoethanoic acid in acid and basic pH. Why do amino acids have relatively high melting points?

Lone pair

Availability of lone pairs

$C_2H_5NH_2$ is also a nucleophile

From CH_3CN, which is not a nucleophile

Amides = $RCONH_2$

P_2O_5 removes H_2O

$LiAlH_4$ removes H

D is acidic

Degradation

E + acid → salt

How many C atoms?

E is water insoluble, F is water soluble

Availability of lone pair

Amine and acid react to form $-N(H)CO-$ link

Acid and base

Zwitterion

10 minutes

Test your knowledge

1 Give a general equation for fragmentation in a mass spectrometer.

2 What information is given by the largest m/z value on a mass spectrum?

3 Why does CH_3Br show molecular ion peaks at m/z 94 and 96?

4 On what principle does infrared spectroscopy rely?

5 Which part of an organic molecule is detected in NMR spectroscopy?

6 What can be deduced from the number of NMR peaks?

7 What can be deduced from the size of NMR peaks?

8 What can be deduced from the splitting of NMR peaks?

9 Why is TMS added to a sample for NMR spectroscopy?

10 What type of structural features are detected by UV visible spectroscopy?

Answers

10 Multiple bonds and delocalised systems
9 As reference peak
8 Number of adjacent protons
7 Number of equivalent protons
6 Number of hydrogen environments
5 Hydrogen-nuclei
4 Vibrating bonds absorb i.r. radiation
3 Presence of ^{79}Br and ^{81}Br isotopes
2 Mr of molecule
1 $[M]^{+}\cdot \rightarrow [X]^{+} + [Y\cdot]$

 If you got them all right, skip to page 79

Organic Spectroscopy

30 minutes

Improve your knowledge

1 Mass spectroscopy gives information about the **carbon skeleton** of organic compounds. The molecular ion, $M^{+\cdot}$, is produced by electron bombardment. This radical **fragments** by breaking single bonds forming a **cation**, which is **detected**, and a **radical**, which is **not**.

$$[M]^{+\cdot} \rightarrow [X]^+ + [Y]^\cdot$$

2 **Fragments** are recorded as peaks on a mass spectrum according to their mass/charge ratio, m/z. Charges are normally +1 so m/z values correspond to the mass of the fragment. The highest m/z value corresponds to the **molecular ion** and so gives the M_r, (molecular mass).

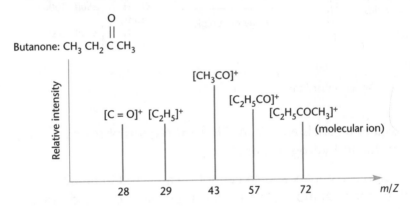

3 Peak **intensity** (height) reflects the fragment **abundance** and the presence of **isotopes**. $^{35}Cl/^{37}Cl$, $^{79}Br/^{81}Br$ and $^{12}C/^{13}C$ have relative abundance of 3:1, 1:1 and 100:1.1. Thus CH_3Cl shows molecular ion peaks at 50 ($[CH_3{}^{35}Cl]^{+\cdot}$) and 52 ($[CH_3{}^{37}Cl]^{+\cdot}$) with relative intensities 3:1. Equally, when a large molecule fragments the most stable fragments will show as the most abundant on the spectra.

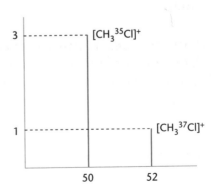

4 Infrared (IR) spectroscopy gives information about **functional groups**. Different **bonds vibrate** at different frequencies and so **absorb different frequencies** of IR radiation. An IR spectrum shows which frequencies are absorbed (in cm^{-1}) as 'downwards' peaks. Bonds and thus functional groups are identified from a **table**. The 'spectrum' below is not for a real compound but is merely a diagrammatic representation of the position and shape of some important absorptions.

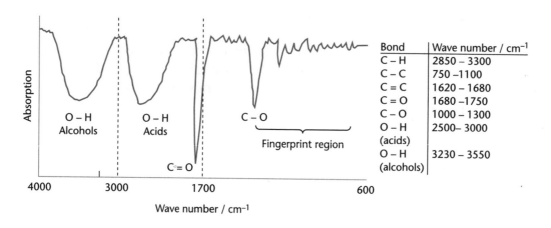

Bond	Wave number / cm^{-1}
C – H	2850 – 3300
C – C	750 – 1100
C = C	1620 – 1680
C = O	1680 – 1750
C – O	1000 – 1300
O – H (acids)	2500 – 3000
O – H (alcohols)	3230 – 3550

O–H bonds in acids and alcohols absorb in **different** regions of the spectrum, but both are broad due to **hydrogen bonding**.

The **fingerprint** region is complex and so cannot be analysed in detail. However, it is unique for each molecule and so can be used to help distinguish compounds, particularly those which have similar absorptions in other areas of the spectrum.

5 **Nuclear magnetic resonance spectroscopy (NMR)** gives information about **hydrogen atoms**. Hydrogen nuclei (protons) have a magnetic field that aligns with an external magnetic field (parallel) – a **low-energy** state. Irradiation with radio waves cause some nuclei to 'flip' to an anti-parallel, **high-energy** state, where their fields are **against** the applied field.

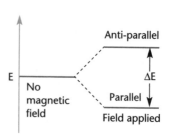

6 Protons in different **environments (non-equivalent)** absorb different frequencies of radio waves, giving rise to **separate peaks** on NMR spectra, at different **chemical shifts**, δ. Non-equivalent protons are those in **different groups** (e.g. CH_3 and CH_2) or in the same group in a **different position** in the molecule (e.g. a terminal CH_3 and a side-chain CH_3). **Aromatic** protons tend to show up as the **same environment**.

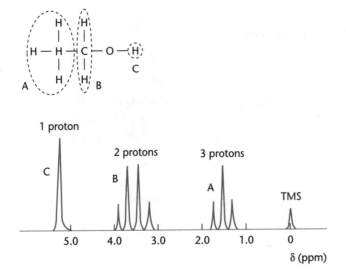

Type of proton	δ (ppm)
$R - CH_3$	0.7 – 1.2
$R_2 - CH_2$	1.2 – 1.4
$R_3 - CH$	1.4 – 1.6
$RCOCH_3$	2.1 – 3.6
$RCOCH_3$	3.1 – 3.9
$RCOOCH_3$	3.7 – 4.1
ROH	0.5 – 5.0
C6H5 – H	7.3
RCH O	9.7
RCO2 H	11.0 – 11.7

7 The **total number** of equivalent protons is shown by the **relative size** of the peak (**integration value**).

8 **High-resolution NMR** shows **spin–spin splitting**. A peak for a particular environment is split by n **adjacent** protons into $n + 1$ peaks. Thus 1 adjacent proton produces a doublet, 2 adjacent protons produce a triplet and 3 adjacent protons produce a quartet. **Aromatic** protons tend **not** to split each other or be split by **other** protons.

9 δ **values** are recorded relative to a **TMS reference peak**. TMS (tetramethylsilane, $(CH_3)_4Si$) is used as a reference because almost all shifts occur to its left (downfield) on a spectrum; it gives a strong, single peak; and it is inert.

10 Ultraviolet (UV) / visible absorption spectroscopy is used to identify the presence of multiple bonds and delocalised systems. The electrons in these systems are able to undergo transitions from one energy level to another by absorption of UV or visible light. The frequency of light absorbed depends upon the nature and environment of the electron system. Hence an aromatic ring can be distinguished from a nitrogen–nitrogen double bond, $N=N$.

Make sure you understand the following key points:

mass spectroscopy / fragmentation / m/z value / isotopic abundance / infrared spectroscopy / NMR spectroscopy / chemical environment / relative integration or intensity / spin–spin splitting / TMS / UV–visible spectroscopy

Organic Spectroscopy

30 minutes

Use your knowledge

1 An unknown compound A, C_4H_8O, showed IR absorption at around 1715 cm^{-1}, mass spectrum peaks at m/z = 57, 43 and 29. High-resolution NMR shows a triplet of intensity 3, a quartet of intensity 2 and a singlet of intensity 3. Reduction of A produces B, which shows a broad IR absorption at around 3300 cm^{-1}. B reacts with conc. H_2SO_4 to form two isomeric products, C and D, which show I.R. absorptions at about 1650 cm^{-1}. Low-resolution NMR of C shows four peaks of relative integration values 3:2:2:1.

a) What can be deduced about A–D from the IR data?

b) What can be deduced about A and C from the NMR data?

c) Write an equation to show the formation of the fragment at m/z 57 on the mass spectrum of A and identify the fragments at m/z = 43 and 29.

d) Give the formula and names of A–D.

2 Explain how the isomeric molecules methyl ethanoate, CH_3COOCH_3, and propanoic acid, CH_3CH_2COOH, could be distinguished by using:

a) Infrared spectroscopy.

b) Low resolution NMR.

c) Mass spectroscopy.

Refer to both molecules in your answers.

Use the diagram page 75 to identify functional groups

Number of NMR peaks = number of environments

Intensities = number of protons

Splitting indicates number adjacent protons, n + 1

Match fragments to m/z values

Use all the data to identify the molecules

Identify unique functional groups

How many environments does each molecule have?

Identify a fragment with a unique mass in each molecule

Synoptic Questions

Organic chemistry

1 Remember the relationship between the various classes of compounds. Try to think of what each type of compound can be converted into and what it can be prepared from. Remember to write balanced equations where relevant.

2 When describing how to carry out organic reactions, you must fully identify every reagent with a full name or formula.

3 Reaction conditions, such as temperatures, catalysts, concentrations and solvent, are often vital and must be clearly stated.

4 Remember that some reactions can lead to more than one isomeric product. For example:
- Elimination can form two structurally isomeric alkenes.
- Alkenes can be produced as cis- and trans- isomers.
- Nucleophilic addition to carbonyls and SN_1 reactions can produce racemic mixtures.

5 Where relevant give practical details for reactions such as apparatus, safety precautions and how the final product is separated and purified.

6 When trying to devise a synthetic pathway involving several steps it is often easier to work backwards from the product molecule.

7 When interpreting spectra it is often helpful if you already have an idea of what the molecule might be, e.g. from chemical data.

8 When interpreting spectra identify all relevant peaks (quoting numbers where appropriate, such as chemical shifts). Relate them clearly to the structural features they represent and use this data to draw conclusions about the molecular structure.

9 Try to remember the commercial applications of organic substances and relate them to their structure and chemistry.

10 When drawing mechanisms make sure that curly arrows are used carefully and accurately. Show all structures clearly.

Synoptic Questions

Sample question

Three structural isomers, A, B and C, have the molecular formula C_4H_8O. A and B produce a silver precipitate when warmed with ammoniacal silver nitrate and react with acidified aqueous potassium dichromate(VI) to produce the organic compounds D and E respectively. Both D and E show strong IR absorptions in the region of 2850 to 3000 cm^{-1}, whereas B absorbs strongly in the region of 1680 to 1750 cm^{-1}. The low resolution NMR spectra of E shows three peaks with intensities in the ratio of 6:1:1. B reacts with HCN gas in alkaline conditions to produce the organic compound F. F shows a form of stereo isomerism. A forms a sweet-smelling compound, G, when boiled with methanol in the presence of concentrated H_2SO_4.

a) Use the chemical and spectroscopic data to identify A–G. What kind of stereoisomerism is exhibited by F?

b) Suggest a chemical test to confirm the presence of the functional group present in A and C. Describe what would be observed.

c) Draw the mechanism for the reaction between B and HCN and explain why the isomerism you have stated in a) could not be readily detected.

Answers

a) Formula suggests carbonyls. Silver mirror with ammoniacal $AgNO_3$ indicates A and C are aldehydes. They undergo oxidation with dichromate to produce carboxylic acids. This is confirmed by IR absorptions, which are for O–H (acids). NMR suggests C has only three proton environments. Peak of 6 must be $(CH_3)_2C$–.
Therefore D is $CH_3CH_2CH_2COOH$ (butanoic acid) and E must be $(CH_3)_2CHCOOH$ (2-methylpropanoic acid). Hence A must be $CH_3CH_2CH_2CHO$ (butanal) and C is 2-methylpropanal $((CH_3)_2CHCHO)$. G must be the methyl ester $CH_3CH_2CH_2COOCH_3$ (methylbutanoate). IR absorption confirms carbonyl, as does reaction with HCN.
B must be the ketone $CH_3CH_2COCH_3$ and F is $CH_3CH_2C(OH)(CH_3)CN$ (2-hydroxy-2-methyl-butanitrile). F shows optical isomerism.

b) Add $NaHCO_3$ solution or solid. A colourless gas is evolved, which turns lime water milky.

c) Because the carbonyl is planar the CN^- nucleophile can attack from above and below equally. Therefore an equimolar mixture of both optical isomers is formed. Their optical activities cancel each other out and so there would be no observed rotation of the plane of plane-polarised light.

Inorganic chemistry

1 Remember that elements (and their compounds) in the same group often react in similar ways.

2 Try to remember which compounds and elements tend to undergo redox reactions, e.g. transition-metal species, halogens, lead(IV) and tin(II).

3 You must know the characteristic colours and properties of the elements and compounds on the syllabus.

4 The reactivity of inorganic compounds is often related to their structure and bonding. Thus ionic chlorides tend to dissolve with water, whereas covalent ones often undergo hydrolysis.

5 The acid-base nature of oxides is often important. All metal oxides react as bases; some also behave as acids, making them amphoteric, e.g. Al_2O_3, PbO, BeO. Non-metal oxides tend to react as acids.

6 Transition metals often exist as coloured compounds and solutions and often form complexes. All transition metals form hexaaqua complexes in solution and undergo deprotonation to form coloured hydroxide precipitates with bases. Some dissolve further and many undergo ligand exchange with NH_3, Cl^-, etc.

7 Remember that there are only a limited number of compounds on the syllabus and so try to match what you know to the evidence in the question.

8 Remember that equations must balance in terms of atoms and charges and that state symbols are often important for redox and ionic equations.

9 The chemistry of many inorganic compounds is related to electronic structure, particularly the hydrolysis of chlorides.

10 When interpreting inorganic redox reactions identify which species are oxidised and which are reduced. E° values must be used if they are given in the question.

Sample question

P, Q and R are metals from either the S or P block.

The chloride of P is white solid that reacts vigorously with water to produce a white, hydroxide precipitate, S, and HCl gas. S is observed to dissolve in both dilute H_2SO_4 and dilute NaOH to form solutions containing the metal-containing species T and U. Aqueous solutions of X are found to be inert to oxidising and reducing agents.

The chloride of Q is an oily, volatile yellow liquid that readily decomposes on heating to produce a white solid, V, and chlorine gas. V is sparingly soluble in dilute nitric acid, but the resulting solution produces a white precipitate, W, when aqueous silver nitrate is added. W is soluble in dilute ammonia. Q undergoes hydrolysis with cold water to produce a brown solid, X. When heated with HCl, X reacts to produce chlorine gas, V and one other product.

R is a white solid that readily dissolves in water to produce an aqueous solution. When treated with dilute sulphuric acid this solution produced a thick white precipitate, Y, which is insoluble in excess acid. When heated in a strong flame, R produces a pale-green flame.

a) Use all the chemical data to identify P to Y.
b) Write ionic equations to represent the reaction of precipitate S with dilute H_2SO_4 and NaOH.
c) Write full chemical equations to show the reaction that occurs when the chloride of Q is heated and its reaction with water to form X.

Answers

a) Vigorous reaction of P and formation of HCl suggests hydrolysis and therefore covalent chloride. Hydroxide reacts with both acid and base and is therefore amphoteric. Metal inert to redox reactions and so has only one oxidation state. Suggests P is Al (or Be), the chloride is $AlCl_3$, T is Al^{3+}(aq), S is $Al(OH)_3$, U is $Al(OH)_4^-$.

Liquid state indicates simple molecular substances. Decomposition suggests it is in an unstable state, evolution of Cl_2 suggests Q is being reduced or is in a high state. Precipitate with $AgNO_3$, soluble in NH_3, means W is AgCl and confirms chloride still present in V. V contains Q in a lower oxidation state. Hydrolysis confirms covalent bonding in chloride and that X is probably a solid oxide. Reaction with HCl suggests W is an oxidising agent, which suggests Pb(IV). Therefore, Q is Pb, V is $PbCl_2$, W is AgCl, X is PbO_2.

Solubility of R in water suggests ionic compound. Flame colour confirms R is barium. Precipitate Y is $BaSO_4$.

b) $Al(OH)_{3(s)} + OH^-$(aq) $\rightarrow Al(OH)_4^-$(aq)
$Al(OH)_{3}$(s) $+ 3H^+$(aq) $\rightarrow Al^{3+}$(aq)

c) $PbCl_4 + 2H_2O \rightarrow PbO_2 + 4HCl$
$PbO_2 + 4HCl \rightarrow PbCl_2 + Cl_2 + 2H_2O$

Synoptic Questions

Physical/quantitative chemistry

1 Try to visualise what is happening in descriptions based on practical situations.

2 Always write balanced equations where appropriate.

3 Work out your approach to the calculation. Avoid rushing in.

4 Look carefully at any numbers given, they usually need to be used somewhere.

5 Remember the stages involved in solving quantitative questions.

6 Write down what you are doing. Numbers without words are meaningless.

7 Where appropriate think about what sort of answer you are expecting. For example, in calculating the pH of a weak acid an answer of 10 is clearly wrong!

8 Try doing the calculation in rough first. This will help you and the person marking your script.

9 Check you have answered the question. If you are asked to calculate a mass then don't give your answer as a volume!

10 Remember units.

Synoptic Questions

Sample question

The equation for the reaction of ethanol with ethanoic acid is shown below:

$$C_2H_5OH_{(l)} + CH_3COOH \rightleftharpoons CH_3COOC_2H_{5(l)} + H_2O_{(l)}$$

3.0g of ethanoic acid and 2.3g of ethanol were mixed with a catalytic amount of concentrated sulphuric acid at 100°C for 1 hour. After this time 50 cm³ of 1.0 M sodium hydroxide was added and the resultant mixture titrated with 1.0 M hydrochloric acid. 33.25 cm³ of hydrochloric acid was required.
   Ignore the sulphuric acid in any calculations as it is only there in a catalytic amount.

a) Predict the effect of adding additional concentrated sulphuric acid to the mixture.

b) Calculate the value of K_c for this reaction at 100°C.

c) By considering average bond enthalpies calculate)H for the reaction and use this to explain the effect on the yield of the reaction if it were to be carried out at a temperature greater than 100°C.

Answers

a) As well as being a catalyst sulphuric acid is also a dehydrating agent. The sulphuric acid will remove the water from the reaction and according to Le Chatelier's principle the equilibrium will move to the right hand side i.e. producing more water and hence more ethyl ethanoate.

b) The number of moles of sodium hydroxide added to the equilibrium mixture was 0.05. The number of moles of hydrochloric acid that reacted with the excess sodium hydroxide was 0.03325. Therefore the number of moles of sodium hydroxide that reacted with the ethanoic acid in the equilibrium mixture was (0.05 – 0.03325) = 0.01675. This means the equilibrium number of moles of ethanoic acid was 0.01675. The initial number of moles of ethanoic acid was 3/60 = 0.05 and the initial moles of ethanol was 2.3/46 = 0.05. The number of moles of ethanoic reacting was 0.05 – 0.01675 = 0.03325. This is also the number of moles of ethanol reacting (because of the 1:1 ratio). Therefore the number of moles of ethyl ethanoate and water produced was 0.03325 and the equilibrium amounts of ethanol and ethanoic acid was 0.01675. $K_c = [0.03325]^2/[0.01675]^2$
= 3.94

c) An O–H bond and a C–O bond are broken. An O–H bond and a C–O bond are formed. Breaking bonds requires energy and forming bonds releases energy. Therefore the overall enthalpy change is 0. As a result, increasing the temperature will not cause an increase (or a decrease) in the yield.

Energetics 2

1 a) i) Enthalpy of formation of lithium iodide

ii) Enthalpy of atomisation of iodine (or $\frac{1}{2}$ bond enthalpy of iodine)

iii) Enthalpy of atomisation of lithium

iv) 1st ionisation energy of lithium

v) 1st electron affinity of iodine

vi) Lattice enthalpy of lithium iodide

b) Lattice enthalpy of lithium iodide = -744 kJ mol^{-1}

c) Li$^+$ has a greater charge density than Na$^+$ because the lithium ion has a smaller ionic radius than the sodium ion. This means that the lithium ion will exhibit a greater attraction for the chloride ion than will the sodium ion.

2 a) $Mg^{2+}_{(g)} + 2Cl^-_{(g)} \rightarrow MgCl_{2(s)}$ (or reverse).

b) -2526 kJ mol^{-1}

c) A Born–Haber cycle assumes the substance is 100% ionic.

3 a) As the group is descended the solubility of the sulphates decreases.

b) For magnesium sulphate the hydration enthalpy compensates for the lattice enthalpy; hence magnesium sulphate is soluble. In the case of barium sulphate the value of the hydration enthalpy is less than that of the lattice enthalpy and so the salt does not dissolve.

Thermodynamics

1 a) Reaction is spontaneous if ΔG^θ is ≤ 0. Change from solid state to aqueous ions leads to an increase in entropy, i.e. $\Delta S^\theta =$ +ve. Even though ΔT^θ is positive, which does not favour the reaction, $T\Delta S^\theta > \Delta H^\theta$, therefore ΔG^θ is negative.

b) ΔH^θ is negative because the reaction is highly exothermic, releasing heat. ΔS^θ is positive because the number of moles on the RHS is greater than the number of moles on the LHS. ΔG^θ is negative because both ΔH^θ and $-T\Delta S^\theta$ are negative.

c) Reaction is only spontaneous if $\Delta G^\theta \leq 0$, i.e. $\Delta H^\theta < T\Delta S^\theta$. ΔH^θ is positive and so this is only true if $T > 373$ K.

2 a) $\Delta S^\theta = S(\Delta S^\theta_{products}) - S(\Delta S^\theta_{reactants})$

therefore

$\Delta S^\theta = (205 + 2 \times 152) - (2 \times 133)$

$\quad\quad = 243$ J K^{-1} mol^{-1}

b) $\Delta G^\theta = \Delta H^\theta - T\Delta S^\theta$

$\quad\quad = 124 - 298 \times (243/1000)$

$\quad\quad = 124 - 66.75$

$\quad\quad = 51.6$ kJ mol^{-1}

c) Reaction becomes spontaneous when $\Delta G^\theta = 0$, i.e.

therefore

$T = \Delta H^\theta / \Delta S^\theta$

$T = 124 / (243/1000)$

$T = 510$ K

Chemical Equilibria 2

1 a) Initial number of moles of acid/alcohol = 0.5 Number of moles of acid/alcohol at equilibrium = 0.17 (0.5 – 0.33), $K_c = 3.77$ (no units).

b) i) The equilibrium will move to the LHS (increasing the temperature favours the endothermic reaction).

ii) K_c will decrease as ratio of products/reactants decreases.

2 a) Calculation of partial pressures $P_{SO_2} = 2.5$ $P_{SO_3} = 1.25$ $P_{O_2} = 1.25$, $K_p = 0.2$ atm^{-1}

b) The equilibrium will move to the RHS, favouring the side with the fewer number of moles.

c) A decrease in temperature favours the exothermic pathway, i.e. the equilibrium moves to the RHS causing an increase in K_p.

Acids and Bases

1 a) Negative logarithm to base 10 of the concentration of oxonium ions.

b) 0.60

c) 0.71

d) A weak acid partially dissociates.

e) K_a of 1.26×10^{-5} leading to $[H_3O^+]$ $= 2.30 \times 10^{-3}$ pH = 2.64

f) pH = 5.38

2 a) See page 20.

b) Any strong monobasic base, e.g. NaOH.

c) Phenolphthalein, pK_{In} corresponds to the end point.

Kinetics 2

1 a) Zero order w.r.t I_2, first order w.r.t. CH_3COCH_3, first order w.r.t. HCl (i.e. rate $= k[CH_3COCH_3]^1[HCl]^1$)

b) 1.76×10^{-5} mol^{-1} dm^3 s^{-1}

c) An aliquot of the reaction mixture should be quenched (e.g. by adding ice) and titrated with sodium thiosulphate solution.

2 $2Fe^{3+}_{(aq)} + 2I^-_{(aq)} \rightarrow 2Fe^{2+}_{(aq)} + I_{2(l)}$

Redox Equilibria

1 a) A = Potentiometer (or high resistance voltmeter) B = salt bridge, conc = 1 mol dm^{-3}

b) $Ni_{(s)} \mid Ni^{2+}_{(aq)} \parallel Ag^{+}_{(aq)} \mid Ag_{(s)}$

c) 0.80 V

d) $Ni + 2Ag^{+} \rightarrow Ni^{2+} + 2Ag$

e) i) $Ag^{+}_{(aq)} + e^{-} \rightleftharpoons Ag_{(s)}$ Dilution will move the equilibrium to the LHS, producing more electrons, hence the $E^{\ominus}$ becomes more negative.

ii) $E^{\ominus}_{cell} = E^{\ominus}_{RHE} - E^{\ominus}_{LHE}$ Inserting a more negative value for the LH electrode value causes a more negative $E^{\ominus}_{cell}$

2 a) $2Fe + O_2 + 2H_2O \rightarrow 2Fe^{2+} + 4OH^{-}$ [or $2Fe(OH)_2$]

b) Zinc corrodes in preference to iron because it has a more negative electrode potential.

Periodic Table 2

1 a) $4Na_{(s)} + 2O_{2(g)} \rightarrow Na_2O_{(s)}$, $S_{(s)} + O_{2(g)} \rightarrow SO_{2(g)}$ or $2S_{(s)} + 3O_{2(g)} \rightarrow 2SO_{3(g)}$

b) Sodium oxide is basic and sulphur dioxide (and trioxide) is acidic.

c) Sodium oxide consists of ionic bonding whereas the oxides of sulphur consist of covalent bonding.

d) $MgCl_2 + aq \rightarrow Mg^{2+}_{(aq)} + 2Cl^{-}_{(aq)}$, $PCl_3 + 3H_2O \rightarrow H_3PO_3 + 3HCl$, $PCl_5 + 4H_2O \rightarrow H_3PO_4 + 5HCl$

e) Covalent chlorides are hydrolysed.

f) Aluminium is an amphoteric metal. Hence the reaction of aluminium chloride with water is a mixture of the reactions of both non-metal chlorides and metal chlorides.

2 a) Lead atoms are bigger than carbon atoms hence the lead outer electrons will be more delocalised, hence a greater degree of metallic character.

b) Silicon dioxide is acidic, lead oxide is amphoteric.

c) See page 40.

d) Tin atoms are willing to promote a 5s electron to the 5p orbital, whereas lead is not prepared to promote from the 6s to the 6p, because tin forms more exothermic bonds since it is smaller.

Transition Metals

1 a) Partially filled d orbital; complexes/coloured compounds/catalysts/variable oxidation states.

b) $3d^3$ / $3d^3 4s$ / $3d^8$

c) Mn^{2+} is d^5 half-filled shell which shows a degree of stability, Fe^{2+} is d^6.

d) +3 corresponds to loss of all 4s and 3d electrons leaving full outer shell.

2 A = $[Cr(H2O)6]^{3+}$ hexaaquachromium (III) ion / B = $[Cr(H_2O)_3(OH)_3]$ / C = $[Cr(NH_3)_6]^{3+}$ hexaaminechromium (III) ion / D = $[Cr(H_2O)_3(OH)_3]$ / E = $[Cr(OH)6]^{3-}$.

3 a) Reacts with acids and bases
$[Al(H_2O)_3(OH)_3]$ + OH^- →
$[Al(H_2O)2(OH)_4]^-$ + H_2O / $[Al(H_2O)_3(OH)_3]$ + 3H+ → $[Al(H_2O)_6]^{3+}$

b) iron (II) hydroxide is oxidised to brown iron (III) hydroxide because Fe^{3+} is d^5 and shows a degree of stability.

c) +3 metal ions have higher charge density so more polarising. O–H bonds weakened and easily deprotonated by H_2O to produce H_3O^+ in solution.

4 a) Homogeneous; variable oxidation number.

b) $S_2O_8^{2-}$ + $2Fe^{2+}$ → $2SO_4^{2-}$ + $2Fe^{3+}$ / $2I^-$ + $2Fe^{3+}$ → I_2 + $2Fe^{2+}$ /oppositely-charged ions colliding rather than two anions so lower activation energy.

Haloalkanes 2

1 a) $CH_3(CH_2)_5CH_2Br$, 1-bromoheptane

b) Both haloalkane and OH^- ion present in a single step and so both influence kinetics.

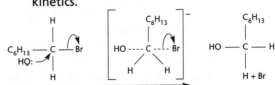

c)

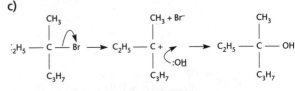

d) B has bulky alkyl groups that prevent easy attack of positive C atom by nucleophile but their inductive effects stabilise a cation intermediate allowing subsequent attack by nucleophile. Hence two steps rather than one.

e) OH^- ions attack planar cation equally from above and below and from both optical isomers in equal amounts, leading to a racemic mixture.

2 a) 1-bromopropane, magnesium metal, dry ether, iodine catalyst.

b)

$$-C-C-C-C-C-C-$$

React propyl magnesium bromide with propanal in dry ether and treat with aqueous HCl to form hexan-3-ol. Then warm with potassium dichromate to oxidise to hexan-3-one.

c) Polar C–Mg bond

d) Because more than one carbon can be added to the chain in a single step. Using cyanide would require many steps which reduces product yield.

Aromatic Compounds

1 a) Six-carbon ring covalently bonded/three pi bonds/delocalised/above and below the ring/all bonds in ring of equal length

b) $C_6H_6 + 3H_2 \rightarrow C_6H_{12}$ /benzene does not release 360 kJ mol^{-1} /due to stability conferred by delocalisation

2 a) $C_6H_6 + HNO_3 \rightarrow C_6H_5NO_2 + H_2O$ /conc. H_2SO_4 / conc. HNO_3 / 50–60°C

b) Electrophilic substitution/addition doesn't occur as this would permanently disrupt delocalisation, reducing stability.

c) $C_6H_5NO_2 + 6[H] \rightarrow C_6H_5NH_2 + 2H_2O$ / Sn / HCl

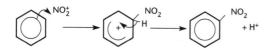

Acids, Ester and Acid Chlorides

1 a) N = propanol / O = sodium propanoate / P = methyl propanoate reduction/ neutralisation/esterification

b) CO_2 / pass through lime water to form cloudy ppt.

c) $C_2H_5COOH + PCl_5 \rightarrow C_2H_5COCl$ (Q) + $POCl_3$ (R) + HCl (S)

d) Propanoic acid CH_3CH_2COOH

2 a) A = $C_2H_5COCH_3$ methyl ethanoate / B = $HCOOC_3H_7$ propyl methanoate / C = $CH_3COOC_2H_5$ ethyl ethanoate / E = methanol / F = methanoic acid / G = sodium ethanoate

b) $HCOOC(CH_3)HCH_3$ /methanoic acid + propan-2-ol

3 a) $CH_3COCl + H_2O \rightarrow CH_3COOH + HCl$ hydrolysis

b) C bonded to electronegative O and Cl so very δ+ so strongly attracts nucleophiles.

c) $C_6H_5COCl + CH_3NH_2$ substituted amide

$$CH_3-\overset{O}{\underset{}{C}}-Cl \rightarrow CH_3-\overset{:O^-}{\underset{}{C}}-Cl \rightarrow CH_3-\overset{O+Cl}{\underset{}{C}}-O-H \rightarrow CH_3-\overset{O}{\underset{}{C}}-OH + H^+$$

:OH$_2$

$$\underset{H \quad H}{\overset{O}{/\overset{+}{\backslash}}}$$

$H^+ + Cl \rightarrow HCl$

d) $(CH_3CO)_2O + CH_3OH \rightarrow CH_3COOCH_3 + CH_3COOH$

Carbonyls

1 a) $CH_3CHO + HCN \rightarrow CH_3C^*H(OH)CN$ / 2-hydroxypropanenitrile

b) Nucleophilic addition

c) Carbonyl is flat so CN^- nucleophile can attack equally from above and below, leading to a 50:50 mixture (or racemic mixture) of both isomers which cancel each other out.

2 a) A = butan-1-ol/B = butanal/C = butanoic acid/
D = Ag metal/E = butanoic acid/
F = 2-hydroxypentanitrile

b) Oxidation/purple to orange to green

c) G = butan-2-one/H = butan-2-ol

d) Add 2,4-dinitrophenylhydrazine solution which forms an orange/red ppt./recrystallise from ethanol filter under suction, dry, take melting point, compare with values in data book.

Organic Nitrogen Compounds

1 a) Lone pair on N accepts protons/
$CH_3NH_2 + H^+ \rightarrow CH_3NH_3^+$

b) Positive inductive effects from alkyl groups make lone pair more available/electron withdrawal by ring makes lone pair less available.

c) $C_2H_5Br + NH_3 \rightarrow C_2H_5NH_2 + HCl$
ethylamine is a nucleophile due to lone pair so further substitution occurs.

d) Reduction of ethanitrile with $LiALH_4$ /
$CH_3CN + 4[H] \rightarrow CH_3CH_2NH_2$ /no further reaction

2 a) A = CH_3CONH_2 / B = CH_3CN dehydration /
C = $CH_3CH_2NH_2$ reduction / D = CH_3CO_2H hydrolysis / E = CH_3NH_2 / F = CH_3NH_3Cl acid-base

b) Hofmann degradation/carbon chain reduced by one.

c) Amine is insoluble so suspension, amine salt has ionic bonding so soluble.

3 a) Lone pair delocalised onto O so not available for bonding.

b) $-[-N(H)CO-C_6H_4-]-$
condensation/polyamide

c) $HOOCH_2NH_3^+ \; H_2NCH_2COO^-$ /amino acid exists as Zwitterion so contains strong ionic bonds.

Organic Spectroscopy

1 a) A contains C=O/B contains OH alcohol/
   C and D contain C=C, alkenes

   b) A contains CH_3 adjacent to CH_2 / CH_2
   adjacent to CH_3 / CH_3 uncoupled C
   contains four environments of CH_3 / CH_2 /
   CH_2 / CH

   c) $[CH_3CH_2COCH_3]^+$ $m/z = 72 \rightarrow$
   $[CH_3CH_2CO]^+$ $m/z = 57 + [CH_3]$ / $m/z = 43$
   due to $[CH_3CO]^+$ / $m/z = 29$ due to $[C_2H_5]^+$

   d) A = $CH_3CH_2COCH_3$, butan-2-one /
   B = $CH_3CH_2CH(OH)CH_3$, butan-2-ol /
   C = $CH_3CH_2CH=CH_2$, but-1-ene /
   D = $CH_3CH=CHCH_3$, but-2-ene

2 a) Acid shows broad absorption between
   2500–3000 cm^{-1} /ester does not

   b) Acid shows three peaks of relative
   intensities 3:2:1 /ester shows two peaks
   of relative intensities 1:1

   c) e.g. acid shows unique peak at $m/z = 57$
   due to $[CH_3CH_2CO]+$ (OR $[COOH]+$ at m/z
   $= 45$, $[C_2H_5]^+$ at $m/z = 29$, $[OH]^+$ at $m/z =$
   17 etc.) /ester shows unique peak at m/z
   $= 31$ due to $[CH_3O]+$ (OR at $m/z = 59$ due
   to $[COOCH_3]^+$, $m/z = 43$ due to $[CH_3CO]^+$)

Notes

Notes

Notes